International Trade Fairs and Inter-Firm Knowledge Flows

Rachael Gibson

International Trade Fairs and Inter-Firm Knowledge Flows

Understanding Patterns of Convergence-Divergence in the Technological Specializations of Firms

Springer

Rachael Gibson
Eval. Ctr. for Complex Health Interventions
University of Toronto
Toronto, ON, Canada

ISBN 978-3-031-20556-9 ISBN 978-3-031-20557-6 (eBook)
https://doi.org/10.1007/978-3-031-20557-6

This Springer imprint is published by the registered company Springer Nature Switzerland AG
The registered company address is: Gewerbestrasse 11, 6330 Cham, Switzerland

For Mom and Dad

Foreword: How to Connect to the World?

How do firms connect with customers, suppliers, and technology partners when they want to expand to new markets abroad? For small and medium-sized firms with little prior experience, and even for multinational firms, this is not a trivial task. There are large risks and uncertainties these agents must face when crossing into sometimes quite different or unknown territory. Not only are languages often different, but firms also step into different cultural contexts with different expectations and patterns of human interaction and, importantly, deviating institutional settings that shape national production contexts and innovation systems (Lundvall, 1992). These differences need to be recognized, understood, and bridged by firms to be able to exploit the opportunities that arise from internationalization (Bathelt et al., 2018).

Scholarly work since the early 2000s has revealed that trade fairs play a crucial role when firms access foreign markets, exchange knowledge over distance, and engage in direct investments abroad (Bathelt & Schuldt, 2005). Trade fairs have existed for centuries (Allix, 1922; Braudel, 1979): They were thriving since medieval times in Europe and developed into their current format as sample fairs with the industrial revolution in the nineteenth century. Especially the most important, large trade fairs generated opportunities for producers and traders to sell products in markets far away from the sites of production, even in different countries. Naturally, these events became a central object of study for business and management scholars to explore the strategies and practices of how firms could effectively advertise their products to customers, extend market penetration, or gain new market access (Bonama, 1983; Bello, 1992). During the 2000s, economic geography and industrial marketing scholars extended this analytical focus to a broader knowledge-based perspective of trade fairs (Borghini et al., 2004; Maskell et al., 2006). Trade fairs were now conceptualized as temporary clusters since the nature of knowledge flows resembles those in permanent industry clusters (Bathelt, 2004). International flagship events or hub shows, in particular, gather all sorts of producers, traders, suppliers, users, and third parties (e.g., media and industry observers) for a few days within the small space of an exhibition center to engage in intensive exchanges surrounding the exhibited artifacts and the contexts within which they are produced and used (Callon, 2017). As such, these events develop into a microcosm of the corresponding industry or technology field (Rosson & Seringhaus, 1995). Producers meet with suppliers and users to engage in vertical knowledge exchanges by discussing ways to improve cooperation, and competitors observe each other's product portfolios and new developments to find inspiration for future innovation processes. And since actors from many different countries are gathered at these events, attendees discuss institutional particularities and differences between countries and market trends abroad.

These knowledge exchanges are very effective due to the co-presence of actors and the specific knowledge ecologies or global buzz that develops during trade fairs and supports collective sensemaking (Maskell et al., 2006). Attendance at these events often generates substantial costs, and participants need to dedicate a substantial amount time to go to these events. Within the exhibition center, trade fair participants get together with agents that represent the entire value chain and originate from different parts of the world. This triggers a myriad of opportunities for planned and unplanned conversations which would not be possible during regular day-to-day operations of the firms. Face-to-face communication and opportunities for observations provide manifold options to acquire information about new products, markets, and potential partners in multiplex meetings. Altogether, this enables firms to shift from selling to relationship marketing during these events (Hansen, 2000; Rinallo et al., 2017). Since the communities that get together during trade fairs are intersecting and are characterized by knowledge bases that are partially overlapping and partially complementary, many possibilities exist to engage in meaningful learning processes.

A large part of recent research has focused on investigating the nature of interactions and knowledge transfers at these events, yet since data about trade fairs is limited, most of this work presents case studies about the patterns of interactions and knowledge flows, the roles of different attendee groups and trade fair organizers, the recurrent nature and dynamics of these events and how they impact their field (Bathelt et al., 2014). While this research has come a long way, it still appears like a patchwork blanket with sometimes limited or anecdotal evidence and many remaining open questions. One crucial question that has been largely unanswered is: *how* do the learning processes of firms take place during these events? There is hardly any information about the technological search processes of firms and how actors acquire knowledge at trade fairs, or the factors that support or hinder them to engage in corresponding learning processes.

It is at this point where the book by Rachael Gibson makes a major contribution to the literature and our understanding of the field. This is the first study that systematically analyzes technological search processes and interactions of trade fair attendees in a mix of small-n and large-N empirical analyses. The investigation is international in character, investigates a cross-sectoral setting, and uses a transdisciplinary approach. This is in many ways a pioneering study as it provides evidence that firms' knowledge acquisition processes are not only dependent on their internal capabilities, but that these processes are also structured by influences outside the firms' boundaries, namely the institutional conditions of the political economy within which they operate. It is well known that institutional conditions differ widely between countries in terms of industrial relations, corporate governance, access to finance, skill production, inter-firm collaboration, and intra-firm organization (Nelson, 1993; Hall & Soskice, 2001). These differences generate distinct types of political economies that are characterized by differences in how firms interact, how they produce and innovate, who engages in learning, how suppliers and customers are involved, and what types of products are being produced. Rachael Gibson's book is one of the first to provide convincing evidence on how these institutional differences affect interaction and search processes at these events. The findings not only suggest that the nature of trade fairs differs between countries due to national priorities (Bathelt & Zeng, 2015), but that the institutional settings within political economies affect firms' actions and interactions when they attend these events. The clarity of the findings in the book speaks to the painstaking and rigorous empirical process underlying this research that enabled Rachael Gibson to overcome difficulties of acquiring reliable data about trade fairs and to consider the large variety of strategies firms apply and the ways in which multinational firms may operate in foreign markets like domestic players. I hope many readers from academia, policy, and the trade fair universe will be fascinated by the storyline Rachael Gibson develops and leads us through in presenting her work.

Last but not least, the question arises about what the role of trade fairs will be in a future world after the COVID-19 pandemic and in the context of a potentially newly developing world order forced by political ruptures, populist governments and, anti-globalization policies worldwide. The recent pandemic has clearly shown that trade fairs are not only prime places of global knowledge exchanges about the dynamics in an industry or technology field, but they can just as much become super-spreader events for contagious diseases and thus need to be re-evaluated from a public health perspective. Furthermore, spearheaded by the Trump administration in the United States between 2017 and 2021, many governments have established trade barriers and policies to prioritize domestic over foreign producers with the political promise to support their national economies (while securing electoral success among an increasingly polarized population).

Where does this leave international trade fairs? Is there even a future for these events? I suggest to exercise care in drawing hasty conclusions. Many have predicted fundamental changes in how people will interact and operate virtually in the future and replace conventional get-togethers. However, it is also clear that knowledge generation and learning processes are more effective in co-present configurations and it is unclear how organizational integrity could possibly be maintained for long or coherence be generated through remote interaction alone (Bathelt & Li, 2022). I am not sure how international trade fairs could become obsolete in the future under these conditions. And I am also doubtful that these events could be held in an online format. Past experiences show that few virtual trade fairs have excelled or survived over a longer time period. It seems the lack of co-presence and face-to-face interaction in virtual settings generates barriers for meaningful exchanges and interaction. I therefore expect that international trade fairs will remain a fundamental exchange forum in any kind of future world order. It is an important feature of

our world that knowledge pools and knowledge dynamics fundamentally differ between distant regions (due to distinct agglomeration effects and industrial specialization) and also between countries' political economies (related to their specific institutional settings). These differences in the knowledge landscape correspond with a vast reservoir of economic potential for firms to innovate and strengthen their competitiveness. It is simply unconceivable that firms will not try to tap into these opportunities as a crucial part of their future business strategies. Therefore, I believe that trade fairs will remain an integral element of the global economic fabric, and Rachael Gibson's work in this book presents us with tools to better understand their roles within and across political economies.

University of Toronto
Toronto, ON, Canada

Harald Bathelt

References

Allix, A. (1922). The geography of fairs: Illustrated by old-world examples. *Geographical Review*, *12*, 532–569.

Bathelt, H. (2004). Toward a multidimensional conception of clusters: The case of the Leipzig media industry, Germany. In D. Power & A. J. Scott (Eds.), *Cultural industries and the production of culture* (pp. 147–168). Routledge.

Bathelt, H., & Schuldt, N. A. (2005). *Between luminaries and meat grinders: International trade fairs as temporary clusters*, SPACES 2005-06, Marburg: www.spaces-online.com.

Bathelt, H., & Zeng, G. (Eds.) (2015). *Temporary knowledge ecologies: The rise of trade fairs in the Asia-Pacific Region*. Edward Elgar.

Bathelt, H., & Li, P. (2022). The interplay between location and strategy in a turbulent time. *Global Strategy Journal*, 12: doi: https://doi.org/10.1002/gsj.1432, forthcoming.

Bathelt, H., Golfetto, F., & Rinallo, D. (2014). *Trade shows in the globalizing knowledge economy.* Oxford University Press.

Bathelt, H., Cantwell, J. A., & Mudambi, R. (2018). Overcoming frictions in transnational knowledge flows: Challenges of connecting, sense-making and integrating. *Journal of Economic Geography*, *18*, 1001–1022.

Bello, D. C. (1992). Industrial buyer behavior at trade shows: Implications for selling effectiveness. *Journal of Business Research*, *25*, 59–80.

Bonama, T. V. (1983). Get more out of your trade shows. *Harvard Business Review*, *61*, 75–83.

Borghini, S., Golfetto, F., & Rinallo, D. (2004). *Using anthropological methods to study industrial marketing and purchasing: An exploration of professional trade shows*, presented at the Industrial Marketing Purchasing Conference, Copenhagen. Online. Available HTTP. http://www.impgroup.org/paper_view.php?viewPaper=4505 (11 May 2022).

Braudel, F. (1979). *Le jeux de l'exchange: Civilisation matérielle, économie et capitalisme, xv–xviii siècle (Civilization and Capitalism, 15th-18th Century)*. Colin.

Callon, M. (2017). Markets, marketization and innovation. In H. Bathelt, P. Cohendet, S. Henn & L. Simon (Eds.), *The Elgar companion to innovation and knowledge creation* (pp. 589–609). Edward Elgar.

Hall, P. A., & Soskice, D. (Eds.) (2001). *Varieties of capitalism: The institutional foundations of comparative advantage*. Oxford University Press.

Hansen, K. (2000). From selling to relationship marketing at international trade fairs. *Journal of Convention and Exhibition Management*, *2*, 37–53.

Lundvall, B.-Å. (Ed.) (1992). *National systems of innovation: Towards a theory of innovation and interactive learning*. Pinter.

Maskell, P., Bathelt, H., & Malmberg, A. (2006). Building global knowledge pipelines: The role of temporary clusters. *European Planning Studies*, *14*, 997–1013.

Nelson, R. R. (Ed.) (1993). *National innovation systems: A comparative analysis*. Oxford University Press.

Rinallo, D., Bathelt, H., & Golfetto, F. (2017). Economic geography and industrial marketing views on trade shows: Collective marketing and knowledge circulation. *Industrial Marketing Management*, *61*, 93–103.

Rosson, P. J., & Seringhaus, F. H. R. (1995). Visitor and exhibitor interaction at industrial trade fairs. *Journal of Business Research*, *32*, 81–90.

Acknowledgements

This study was first prepared as a PhD dissertation at the University of Toronto. New material, particularly related to more recent developments affecting the global political economy, has since been added. This project has taken me a long time to complete, and there are many people to thank for their help along the way. First, and foremost, my dissertation supervisor, Professor Harald Bathelt, has been an extraordinary source of support and guidance from the initial conceptualization of this project, through the field research stage, to the actual writing. I have gained so much from his deep knowledge of the topic, his detailed and insightful feedback on my work at various stages, as well as his warmth and generosity. This book would neither have been started nor completed without him. It has been a privilege and a great pleasure to work with Harald for so many years. This research also greatly benefited from the thoughtful comments and suggestions of Professors Louis Pauly, David Wolfe, Rod Haddow (all from the University of Toronto), and Peter Maskell (from the Copenhagen Business School). Many thanks to each of them for their support and guidance.

I extend my sincere gratitude to several dear friends and colleagues for the various ways in which they have supported me during the years spent researching and writing this book. I look forward to thanking each of you personally over a cup of tea or glass of wine in the near future. I would also like to say a special thank you to Zachary Romano, senior publishing editor at Springer, for his enthusiastic support of this project as well as his expert and timely guidance throughout the entire publication process. I am also extremely grateful to the manuscript's two anonymous reviewers whose insightful comments and feedback helped strengthen the finished product.

The study's empirical analysis is based on the many trade fair exhibitors who agreed to be interviewed or completed a questionnaire. I am so appreciative of their time and the insights they shared with me; this book would not have been possible without them. I am also grateful for the financial assistance I received from the Department of Political Science at the University of Toronto, the Social Sciences and Humanities Research Council, and the Ontario Graduate Scholarship during the early stages of my research and writing.

Some ideas and conceptualizations developed in this research have been published elsewhere. Many thanks to Taylor and Francis Ltd. on behalf of Regional Studies Association for granting permission to use material from my 2015 article, co-authored with Harald Bathelt, entitled "Learning in 'Organized Anarchies': The Nature of Technological Search Processes at Trade Fairs," *Regional Studies* 49(6): 985–1002. Thanks also to Edward Elgar Publishing Limited for granting permission through PLSclear to include ideas from my chapter, co-authored with Harald Bathelt, "Proximity Relations and Global Knowledge Flows: Specialization and Diffusion Processes Across Capitalist Varieties," in *Regional Development and Proximity*, eds. André Torre and Frédéric Wallet (2014).

Finally, I owe so much to my family. To my loving partner, Mike, and our beautiful and brilliant son, Griffin, thank you for helping create the time and space I needed to complete this project, for nurturing me with good food and conversation, and for keeping my spirits up with your thoughtfulness, hugs, and sense of fun. I would also like to thank my sister, Sarah, and my brother, Lynn, and their families, for being there for me in so many different ways for as long as I can remember. A special thank you also to my mother-in-law, Jennie, for her kindness and care. Lastly, words cannot fully express how much I appreciate and have benefited from the unwavering support, encouragement, and love of my parents, Janice and Lynn Gibson. Their values, work ethic, and generosity are a constant source of inspiration to me. It is with great pleasure that I dedicate this book to them.

Though I have greatly benefited from the help of all those mentioned above, responsibility for errors of fact or interpretation is, of course, mine alone.

Contents

Abbreviations

AUMA	Association of the German Trade Fair Industry
B2B	Business-to-Business Trade Fair
B2C	Business-to-Consumer Trade Fair
CC	Comparative Capitalisms
CES	International Consumer Electronics Show
CME	Coordinated Market Economy
CMTS	Canadian Manufacturing Technology Show
EG	Economic Governance
EMO	Exposition Mondiale de la Machine-Outil (Machine Tool World Exposition)
F2F	Face-to-Face
LFI	LightFair International
LME	Liberal Market Economy
ML	Medium-Sized or Large Firm
MMTS	Montreal Manufacturing Technology Show
NBS	National Business Systems
NPE	International Plastics Showcase
NSI	National Systems of Innovation
RE	Relative Evaluation
SSP	Social Systems of Production
TSEA	Trade Show Exhibitors Association
VoC	Varieties of Capitalism

List of Tables

Chapter 1
The Role of International Trade Fairs in the Global Political Economy

1.1 Introduction to the Study

The past decade or so has been marked by events that have called into question the era of international trade and commerce that defined the early 1990s to 2010 – the so-called "golden age" of globalization or the period of "hyperglobalization" (The Economist, 2019a; Rodrik, 2011; Subramanian & Kessler, 2013). The 2008–2009 financial crisis sparked serious doubts about the advantages of global interconnectedness, even amongst business leaders (Ignatius, 2017). And, by 2016, events like the Brexit referendum in the United Kingdom and the election of Donald Trump as President of the United States were cited as illustrative examples of a broader rise in protectionist economic policies and nationalist movements taking place around the world (Goodhart, 2017; van Bergeijk, 2019; Flew, 2020).[1] Three years later, *The Economist* observed that "globalisation has given way to a new era of sluggishness", using the term "slowbalisation"[2] to capture the slow-down of global trade flows over the past decade and the seeming fragility of the global economic system (The Economist 2019a, 2022). Now, in 2022, this "sense of retreat from globalization" (Flew, 2020) has coincided with new forces – first, the ongoing COVID-19 pandemic and the accompanying disruptions to global supply chains; second, the war in Ukraine and the economic sanctions that have followed – which further push

[1] While it appears that the number of populist leaders in power, including Donald Trump, has declined since the beginning of 2021 (See Meyer, 2022), this has not necessarily meant an immediate or clear reversal of the protectionist economic policies that were put in place by previous administrations. For instance, several commentators have noted that the Biden administration's trade policy has not significantly departed from that of the former Trump administration – though it has clearly been a quieter, more "polite" continuation of Trump's style of protectionism (Bacchus, 2022; Frum, 2022).

[2] This term was coined, in 2015, by Dutch writer and trendwatcher Adjiedj Bakas to characterize the observed backlash against economic globalization (The Economist, 2019b; Jones, 2020).

R. Gibson, *International Trade Fairs and Inter-Firm Knowledge Flows*,
https://doi.org/10.1007/978-3-031-20557-6_1

towards a reimagined form of global economic organization (Baschuk, 2022; The Economist, 2022). All of this has introduced considerable uncertainty regarding the future of the global political economy and raises important questions about how firms can best respond and adapt to a rapidly changing global economic landscape.

It is against this backdrop of political-economic uncertainty that the present study develops the argument that international trade fairs can play a critical role in supporting national economies by providing firms with, for example, opportunities to gain access to new markets, network with key actors in their particular industry or value chain, and tap into valuable external knowledge flows regarding new technologies and innovations (Bathelt et al., 2014). More centrally, this study suggests that international trade fairs have become crucial nodes in the global political economy, driving global economic dynamics and shepherding the negotiation of differences between capitalist economies with respect to their technological and institutional practices and conditions. In so doing, I argue that international trade fairs represent a decisive mechanism by which distinct national patterns of technological specialization may converge or diverge. Through the development of these arguments, this book underscores the potential significance of these events in relation to key debates in the study of comparative political economy, specifically, and the field of political science more broadly.

International trade fairs have been largely neglected in political science. Yet, in an era commonly characterized by the increasing integration of economic linkages on a global scale, these events have become important – albeit temporary – platforms for establishing trust with geographically distant business partners and developing durable inter-firm networks in areas such as production, research, and marketing (Borghini et al., 2004; Maskell et al., 2006; Bathelt & Schuldt, 2008). By bringing together prominent as well as less familiar actors from a particular industry, value chain, or technology field, trade fairs enable the exchange of industry "best practices" and information about technological advancements, customer preferences, and related adjustments in institutional or organizational practices. Moreover, because these events are typically organized according to a specific industry or technological focus, international trade fairs have the capacity to cut across traditional geographical and political boundaries, thereby creating multiple opportunities for face-to-face (F2F) interaction and knowledge exchange among firms from different national political economies.

This study is situated within the context of the broader Comparative Capitalisms (CC) literature (Deeg & Jackson, 2007; Jackson & Deeg, 2008),[3] and particularly Hall and Soskice's (2001) seminal work, Varieties of Capitalism (VoC), which

[3] This study adopts the term "comparative capitalisms" (CC), which was originally used by Jackson and Deeg (2006) to characterize a number of approaches – perhaps most notably the *Varieties of Capitalism* – that can be distinguished from other comparative approaches to the study of political economy. While the present study draws on Jackson and Deeg's work specifically, the term "comparative capitalisms" has since become widely used (see, for example: Ebenau et al., 2015). Chapter 3 of this book provides a fuller description of the CC literature and how it is used to inform the present study.

dominated the field of comparative political economy for more than a decade following its publication in 2001 (Bruff et al., 2015). Within this intellectual tradition, I focus specifically on the realm of *inter-firm relations* – which, of the five main institutional spheres identified in Hall and Soskice's work (2001),[4] has received the least academic attention. In adopting this focus, my study aims to identify and better understand the processes by which new technologies, organizational norms and practices, and their associated knowledge bases may be exchanged among firms and, ultimately, across different capitalist varieties. One of the principal objectives of the research is to clarify the role of international trade fairs in enabling inter-firm interaction and knowledge flows across national political economies and arrive at some reasoned expectations as to how such processes might influence future production patterns in relation to national technological specialization. The study is driven by the following main research questions:

1. What is the potential significance of international trade fairs in the context of the global political economy and in relation to recent theories regarding the nature and dynamics of institutional change – particularly those involving endogenously-driven and more incremental forms of institutional change?
2. What is the precise nature of the interaction processes of exhibiting firms during trade fairs, specifically with respect to knowledge flows related to new technological innovations in a given industry or sector?
3. How might inter-firm interactions at trade fairs impact future patterns of technological production, first, at the level of the firm but, ultimately, more broadly at the level of regional and national states? Can we expect such patterns to lead towards convergence on a single national model or continued divergence?

These questions are inspired by previous work that has been published in different literatures and fields of inquiry – including, comparative political economy, economic geography, marketing, international business studies, and organizational behaviour. The present study works towards a more integrated or holistic treatment of the above research questions by bringing together the different and often seemingly disconnected literatures from which such questions are derived. With respect to questions around institutional change, for example, and the nature of national level production patterns, the study is specifically guided by the comparative capitalisms literature, which includes, as mentioned above, the VoC approach, but also the National Business Systems (NBS), Social Systems of Production (SSP), and National Systems of Innovation (NSI) approaches. Taken together, these different approaches have sparked my curiosity about the future structure of national patterns of technological specialization and the mechanisms by which such patterns remain stable or shift over time. Both past and continued discussions about the likelihood of cross-national convergence/divergence are of great relevance here (Crouch & Streeck, 1997; Kitschelt et al., 1999; Yamamura & Streeck, 2003). The core

[4]The five spheres focused on in Hall and Sokice's (2001) work are: industrial relations, vocational training and education, corporate governance, inter-firm relations, and employees.

question in this debate is whether distinct political economies, when faced with the various forces of competitive convergence (e.g., market-driven, technological) "must thrust aside their distinctive institutions and the competitive advantages they have yielded in the past or whether they will be able to adapt and retain such institutions and preserve the social cohesion and economic competitiveness of their societies" (Streeck & Yamamura, 2003, 4). Also relevant are discussions within comparative political economy regarding the need for better or more dynamic theories of institutional change, particularly with respect to change that is more gradual and endogenous in nature (see, for example, Thelen, 2003, 2004; Streeck & Thelen, 2005; Deeg & Jackson, 2007; Hall & Thelen, 2009; Mahoney & Thelen, 2010; van der Heijden & Kuhlman, 2017).

At the micro-level, the focus of this research is on the actual interactions of exhibiting firms during international business-to-business (B2B) trade fairs. On this front, the study's research questions are largely guided by previous work in the field of organizational behaviour. This literature includes the seminal work of Cohen et al. (1972) on "organized anarchies" and "garbage-can" processes of decision-making, but also specific theoretical and empirical investigations into the nature of firms' search behaviours (e.g., Cyert & March, 1963; Gavetti & Levinthal, 2000; Rosenkopf & Nerkar, 2001). Beyond these key works, the study also draws inspiration from the rapidly growing subfield of behavioural economics, particularly existing research which focuses on the firm (see, for example, Armstrong & Huck, 2010; Armstrong & Huck, 2014).

Finally, occupying the meso-level of the analysis, recent work from the fields of economic geography (Maskell et al., 2006; Bathelt & Schuldt, 2008) and marketing (Golfetto & Rinallo, 2013, 2014), in particular, have inspired my focus on international trade fairs as key sites of theoretical and empirical relevance in relation to comparative political economy. More specifically, these recent works have encouraged us to think about international trade fairs differently – as designated spaces in which firms from all over the world (and from different models of capitalism) come together to exchange knowledge about their industry and technological innovations. It is through the lens of trade fairs as "temporary clusters" (Maskell et al., 2006) that a strong argument can be made about the potential of these events to initiate or enable a series of incremental adjustments that, over time, may cumulatively lead to a shift in the future development paths of participating firms.

Based on this multi-level conceptual framework, which will be explicated over the next three chapters, my research aims to uncover the nature and dynamics of inter-firm interaction at trade fairs and then connect these processes to the specific institutional arrangements found in distinct capitalist varieties. I argue that trade fairs provide a critical mechanism linking the micro processes of inter-firm interaction to different capitalist varieties and, in creating such linkages, provide unique "windows of opportunity" through which decisions about the future direction of firms' technological development paths may be made. It is through the context of international trade fairs that such linkages are generated, and it is through the trade fair setting itself that such linkages can be systematically explored and analyzed.

My research is driven by three main goals, each corresponding more or less to a different level of analysis. First, drawing on the garbage-can model (Cohen et al., 1972), as well as ideas from the search and behavioural economics literatures, this book aims to develop a deeper understanding of the nature of firms' search and information processes, specifically with respect to the identification of new technologies and innovations – or "solutions". Second, in reference to recent conceptualizations of trade fairs as "temporary clusters" (Maskell et al., 2006; Bathelt et al., 2014), the book builds on existing empirical evidence regarding the significance of these events in knowledge exchange processes by adopting a specific focus on the political-economic nature and implications of such knowledge flows in relation to the development paths of distinct capitalist models. Finally, based on discussions in the comparative capitalisms literature, this study offers some preliminary insights into how the institutional foundations of capitalist varieties might play a role in shaping inter-firm interactions at trade fairs and, in turn, guiding future patterns of technological production.

1.2 Research Design: The Value of Studying Trade Fairs

Trade fairs are an ideal setting in which to explore the main research questions raised in this book because they can act as a decisive mechanism connecting the micro processes of search and inter-firm interaction to the macro institutional structures that characterize distinct political-economic systems. Following the knowledge-based perspective of trade fairs (Bathelt et al., 2014, 279), this study views international trade fairs as "key nodes in the global political economy", fueling processes of inter-firm learning, knowledge exchange, and innovation. As Chap. 2 will explain, international trade fairs bring together key actors from a given industry or technology field for the primary purpose of exchanging knowledge about new trends and innovations in their industry. Trade fair participants are immersed in a context of rich and varied sources and types of knowledge, which often flow across distinct geographic and political-economic lines. The types of knowledge flows that occur at these events have the potential to link markets and technologies with institutional and political conditions. It is difficult to think of other settings in which these kinds of linkages are created in such a direct manner.

At a general level, then, one might reasonably expect trade fairs in the global knowledge economy to initiate the spread of new technologies, which then become globally available at roughly the same price (Maskell & Malmberg, 1999a). Following this to its logical conclusion, international trade fairs would be strongly implicated in the decline of national specialization patterns and the eventual convergence of political-economic systems towards a standard model (Gibson & Bathelt, 2010). Yet, as the forthcoming empirical analysis will demonstrate, the search processes of firms within the "organized anarchy" (Cohen et al., 1972) of the trade fair serve to complicate this interpretation. By linking these search processes with the national institutional arrangements that define the structures and production contexts of firms, this study provides evidence that the actions and interactions of firms at trade fairs generally support knowledge contextualization and localization (Storper, 1997; Asheim, 1999), thereby reproducing or reinforcing national specialization patterns, rather than undermining them.

The trade fair setting also offers unique methodological advantages, particularly when it comes to the collection of the data. For example, trade fairs are spatially and temporally bounded organizations. They usually take place in convention centres on large fair grounds, and run anywhere from 2 to 3 days to as long as 7–10 days. During these designated periods of time, key actors (e.g., exhibitors) from a particular industry, value chain, or technological field are gathered together in a single location. The precise placement of exhibiting firms in the trade fair venue is also clearly mapped out and documented in both online and printed materials, typically well in advance of the event. Taken together, these conditions reduce many of the costs and challenges commonly associated with administering surveys or conducting interviews (Lampel & Shapira, 1995; Lampel & Meyer, 2008).

1.2.1 Knowledge Flows as a Form of Inter-Firm Interaction

Although inter-firm interaction can assume different types, involve different processes, and occur at different spatial scales (Visser, 2009),[5] this study focuses specifically on knowledge flows as a form of inter-firm interaction. This focus fits with the knowledge-based view of trade fairs (Bathelt et al., 2014), as well as knowledge-based theories of the firm and industrial clusters (e.g., Nonaka, 1994; Nonaka et al., 2000; Grant & Baden-Fuller, 2004). In the globalizing knowledge economy, the economic growth and sustainability of firms – along with the regional and national-states in which they operate – relies very much on being "plugged-in" to ongoing knowledge flows about new developments and innovations in a given industry, technology field, or market. As will be described at some length in Chap. 2, international trade fairs have recently been identified as pivotal platforms for such processes of knowledge creation and exchange. What is of central interest here, however, is that these knowledge flows have the capacity to cut across political-economic boundaries in the context of the trade fair, thereby creating multiple opportunities for firms from different capitalist models to obtain knowledge from one another. It is the dynamics and potential implications of such knowledge flows that form the basis of this work.

My focus in this study is on *the nature of inter-firm interaction processes in relation to the different institutional arrangements of capitalist varieties* – namely, Coordinated Market Economies (CMEs), exemplified by the political economy of Germany, and Liberal Market Economies (LMEs), exemplified by the United States.

[5] Visser (2009) refers to different types of inter-firm interaction, such as market interactions around price negotiation, for example, subcontracting arrangements, network relations between different agents requiring external input for innovation, as well as pure spillovers.

While the CME-LME distinction has been criticized for failing to adequately capture the full diversity of capitalist systems in the global political economy (Crouch, 2005; Katzenstein, 2006), it is a distinction that is largely accepted with regards to the prototypical cases of Germany and the U.S. For this reason, I follow the original VoC framework developed by Hall and Soskice (2001) in adopting the classic CME-LME distinction. Moreover, for present purposes, focusing on Germany and the U.S. makes good sense given that both are considered among the most important host countries for international trade fairs (AUMA, 2009; Bathelt & Schuldt, 2008).

By focusing on inter-firm interactions in relation to the institutional foundations of CMEs and LMEs, I am ultimately interested in the nature of technological adaptation processes – i.e., standardization versus specialization – and the extent to which firms from different capitalist varieties appear to be following a pathway towards convergence or divergence in terms of technological production. While I am unable to directly address the convergence-divergence question with the present data, I aim to capture this variable in a latent way by concentrating on the similarities and differences in the core competencies of firms compared to that of the firms they seek out as potential partners. For the quantitative measurement of this variable, two approaches were adopted. The first involves a comparison of the core competencies of firms and their partners separately. This is done by analyzing the core capabilities of firms and the competencies they look for in partners across several domains – i.e., innovation focus, low-cost focus, customized production, changing product line, and R&D focus. The second approach looks at the firm-partner competence relationships together through a series of *difference variables.*[6]

The qualitative approach to studying the convergence-divergence question involved asking firms to describe how they identify problems and solutions at the trade fair, as well as the processes underlying the generation of networks. On the topic of network generation, for instance, I asked questions such as: Who do you interact with during trade fairs? Do you systematically look for new partners during this trade fair? If so, in which area: distribution, production, projects? How do you go about finding partners? Again, while the data derived from these questions do not allow me to establish a causal link between the interaction processes at trade fairs and national-level outcomes regarding standardization/specialization, they do provide clues about the pathways firms may choose with respect to their technological choices. Over time, these pathways may have broader implications at the level of national-states.

In focusing on knowledge flows as a form of inter-firm interaction, this study specifically focuses on the knowledge that is exchanged between exhibiting firms within the context of leading international trade fairs. The concept of knowledge

[6] To create the *difference variables*, each competence scale ranged from complete homogeneity (matching a firm's core competencies and those it looks for in a potential partner) to complete heterogeneity or complementarity (which refers to the extreme degree of divergence between the same two variables). The scale is, thus, essentially the difference between the firm and partner variables, either in one direction (i.e., towards homogeneity) or the other (i.e., towards heterogeneity or complementarity).

flows is both complex and ambiguous. For example, in a review of the recent literature on knowledge processes,[7] Micek (2016, 55) identifies six different knowledge-related terms, which he situates under the broad term "knowledge interactions".[8] There is little consensus regarding the distinction among such terms, with many scholars employing two or more of these terms interchangeably in their work and often within the context of a single paper. Although it may not be problematic to treat some knowledge-related terms as synonymous (e.g., knowledge circulation and knowledge dissemination), some of the terms suggest semantic if not more substantive differences (Micek, 2016, 62). The multiplicity of knowledge terms makes it difficult to compare findings from different studies and across disciplines. Working towards a common terminology is, thus, seen as a helpful step in addressing this challenge. At the very least, it is important to aim for greater clarification regarding what the different knowledge terms actually mean and how they are being measured.

With this in mind, the present study uses the terms *knowledge flows* and *knowledge exchanges* interchangeably and draws from the work of Griliches (1992) and others (see, for instance, Peri, 2005) in considering such processes in broad terms of learning. More specifically, I adopt the view that "[k]nowledge flows occur whenever an idea generated by a certain institution is learned by another institution. These knowledge flows denote a process of learning from someone else's ideas, effectively building up a stock of accessible (or 'borrowed') research and development" (Peri, 2005, 308; see also Griliches, 1992). Drawing from recent work that takes a search perspective on organizational learning and knowledge acquisition, such processes can be defined as "the problem-solving activities that involve the creation and re-combination of technological knowledge" (Katila & Ahuja, 2002, 1184; see also Winter, 1984). In this sense, the exchange of knowledge between firms is largely the result of one such firm searching for a solution to a problem (Weterings & Ponds, 2009). This view seems particularly relevant when investigating inter-firm knowledge processes in the context of trade fairs, as these events are at least partially framed as an opportunity to explore new innovative solutions to existing problems in, for example, production or design.

This study recognizes the multiplicity of knowledge flows and, thus, focuses on a number of different aspects of such flows. In the quantitative portion of the analysis, the following aspects of knowledge flows are explored: (i) Where does the firm acquire knowledge during the trade fair (e.g., customers, competitors, suppliers)? (ii) What types of knowledge does the firm get (e.g., product-related, technology-related)? (iii) To whom does the firm provide knowledge at the trade fair (e.g., customers, competitors)? (iv) What types of knowledge does it provide (e.g.,

[7] Micek's analysis of the literature includes recent articles published in the fields of economic geography, regional and urban studies. The literature review consists of 89 papers published in journals including *Regional Studies*, *Journal of Economic Geography*, and *European and Planning Studies* (2016, 60).

[8] The six frequently used terms are: "knowledge flows", "knowledge spillovers", "knowledge exchange", "knowledge transfers", "knowledge diffusion", "knowledge dissemination", and "knowledge sourcing" (Micek, 2016, 55).

industry-related, market-related)?[9] The qualitative portion of the analysis examines the following aspects of knowledge flows: (i) Do firms actually search for knowledge at trade fairs? (ii) What exactly do they search for? (iii) How do they search?[10]

In addition to the issues of conceptual ambiguity noted above, knowledge flows are also difficult to capture empirically. They are uneven across space and time, take place under widely different conditions, and produce diverse local, regional, and national development paths (Bathelt & Glückler, 2011). Knowledge flows also follow different channels. In the context of trade fairs, for instance, knowledge exchange processes can be characterized according to their horizontal (between firms and competitors) or vertical (between firms and customers or suppliers) dimensions (Bathelt & Schuldt, 2008). Some types of knowledge, often the more codified forms, seem to flow relatively easily across geographical boundaries, whereas others, especially knowledge that is more tacit in nature, have a tendency to remain highly localized and context-specific (Maskell & Malmberg, 1999b; Gertler, 2003). Trade fairs have increasingly been viewed as important sites through which different types of knowledge are circulated among participants. This includes firms and their corresponding knowledge bases that are derived from diverse capitalist models.

When it comes to measuring knowledge flows, traditional indicators, such as patent citations, are often restricted to measuring codified forms of knowledge, as opposed to more tacit or embodied forms of knowledge. They also provide only an indirect measure of knowledge flows. Further, such indicators are unable to capture the complexity of knowledge flows, which, as previously mentioned, often follow multiple and diverse pathways in response to the strategic behaviour of firms (Arundel & Constantelou, 2006). In response to these challenges, survey indicators of knowledge flows have started to gain traction in recent years. The main approach here is to ask firms directly about the ways in which they utilize different sources and types of knowledge. Theoretically, surveys have the advantage of being able to measure a wide range of knowledge types, including tacit and embodied knowledge. In their recent work, Arundel and Constantelou (2006) find that existing survey data are generally quite good at capturing different types of knowledge flows, but they tend to be less informative on questions regarding "the incentives for firms to participate in these flows and the various channels and mechanisms that facilitate them" (Arundel & Constantelou, 2006, 55). In developing surveys, it is therefore important to place greater focus on the channels and underlying mechanisms of knowledge flows, as well as the type of knowledge firms are seeking from particular sources (Arundel & Constantelou, 2006). The survey and interview guide developed for this study incorporate such a focus (See Appendices A and B).

[9] Appendix A contains the full questionnaire used for both online and face-to-face modes of data collection. Among other things, this questionnaire includes the complete set of questions focused on knowledge flows.

[10] Appendix B contains the complete interview guide used in the qualitative analysis, which includes the questions designed to explore the different aspects of knowledge flows occurring in the trade fair setting.

1.2.2 Adopting a Mixed Methods Approach

In response to the limitations of traditional approaches, some scholars have suggested that quantitative studies alone do not have the capacity to address deeper questions about processes of knowledge creation and exchange (Sotarauta et al., 2011). In a similar vein, Bienkowska et al. (2011, 24) find that "a combination of qualitative and quantitative data and analysis is particularly suitable when trying to assess, within the same space-time context, both overall patterns of labour mobility and individual motives and attitudes of central actors towards mobility." Here, labour mobility is used as a proxy for knowledge spillovers. Micek (2016, 66) supports this view, suggesting that "only an integrated approach (mixed methods) allows to correctly capture knowledge interactions." Through a mixed methods approach, the present study aims to illuminate both the precise nature of firms' search and information processes at trade fairs as well as general patterns in the actions and interactions of firms that belong to distinct capitalist varieties. On the first point, the focus is on identifying patterns of exhibiting firms' search and information practices, as well as the underlying rationale behind them. These are investigated through a qualitative analysis based on semi-structured interviews with 65 exhibiting firms from four leading North American trade fairs. These 65 interviews inform the development of an original typology of search processes, which is then used to explore the connections between different kinds of search processes and the development of national technological specialization patterns.

To identify general patterns of inter-firm interaction at trade fairs, I focus on similarities and differences in the strategies, motivations, and knowledge practices of exhibiting firms. I approach these general patterns through a quantitative analysis based on 165 surveys of German and U.S. firms that were collected through two different data collection procedures. The first was an online survey of past exhibitors from four leading international trade fairs – two in the U.S. and two in Germany. The second was a paper survey distributed and collected in person during two leading international trade fairs – one in the U.S. and one in Germany. The sub-sections below provide a more detailed description of the procedures and instruments used for both the qualitative and quantitative components of this study.

1.2.3 Data Collection Instruments and Procedures

Data gathering for this research was conducted in different stages and consists of multiple sources. As such, a fairly detailed description of the procedures and instruments used for both the qualitative and quantitative research components follows below.

1.2.3.1 The Qualitative Study

The qualitative component of the analysis includes semi-structured interviews conducted at leading international trade fairs, participant observation during these events, as well as a critical review of the available texts, materials, and displays created and presented by exhibiting firms. This combination of data sources provides a more detailed picture of the conditions under which firms operate at trade fairs as well as the particular dynamics of technological search at these events. Although most of the qualitative data were collected through semi-structured interviews with exhibiting firms at their booths during the trade fair, participant observation and review of primary documents from exhibitors were also extremely useful. With respect to participant observation, for instance, I was embedded within the trade fair context for approximately 8–10 h per day over a period of 3–5 days, depending on the trade fair. This provided opportunities for first-hand observation of the conditions under which inter-firm knowledge exchanges occur during such events as well as variations in such conditions across the different fairs.

In conducting the semi-structured interviews, I used an interview guide designed to collect data on the same variables and concepts as those targeted in the self-administered questionnaire used for the quantitative study. The interview guide was divided into four separate sections according to the following themes: (i) trade fair participation, (ii) problem and solution identification, (iii) network generation, and (iv) background information. The first theme parallels that of the first three questions asked in the self-administered questionnaire described below. Its main purpose is to identify the role of trade fairs from the perspective of the respondent firm through questions about the number, frequency, and history of trade fair participation, as well as some initial stated objectives for participation. The second and third themes closely correspond to and aim to build on the specific knowledge and firm competencies questions in the quantitative survey. The final section, in addition to gathering basic data on firm characteristics (e.g., age, number of employees, country of origin), asks about the importance of trade fairs compared to other ways of connecting with customers/finding new partners. This line of questioning runs parallel to the question about activities used to obtain industry-related information in the self-administered questionnaire.

Using the interview guide described above, semi-structured interviews were conducted with 216 exhibiting firms over the course of four North American trade fairs – two in Canada and two in the U.S. The rejection rate ranged from a low of 12% at the Canadian Manufacturing Technology Show (CMTS) in Toronto to a high of 30% at the International Plastics Showcase (NPE) in Chicago (see Table 1.1 below). Of the 216 interviews conducted, 65 were specifically focused on the nature of firms' technological search processes.[11] These 65 interviews serve as the

[11] The qualitative portion of this research draws its data from a larger qualitative study that included multiple themes. This larger qualitative project used four different, partially overlapping, interview guides, one of which was specifically focused on technological search and knowledge processes. It is this guide which forms the basis of the present qualitative analysis.

Table 1.1 Interview statistics of the four North American trade fairs studied in the qualitative analysis, 2009/2010

Trade fair	Interviews conducted		Rejections		Interviews focused on search processes	
	Number	Share (%)	Number	Rate (%)	Number	Share (%)
LFI, New York	55	26	12	18	17	26
NPE, Chicago	69	32	30	30	19	29
CMTS, Toronto	57	26	8	12	17	26
MMTS, Montreal	35	16	7	17	12	19
Total	216	100%	57	26	65	100

Notes: *LFI* LightFair International, *NPE* International Plastics Showcase, *CMTS* Canadian Manufacturing Technology Show, *MMTS* Montreal Manufacturing Technology Show

foundation of the present study's qualitative analysis. Each semi-structured interview lasted, on average, between 10 and 15 min. Given the bustling nature of the typical trade fair, 10–15 min is a fairly generous time commitment.

The focus of this study is on the interactions of firms, rather than on the individuals being interviewed. Thus, in selecting potential interviewees, I sought to speak directly with firm executives, owners, or senior managers, all of whom have a greater capacity to provide information from the perspective of the firm itself. While this was relatively easy to accomplish in the case of small- and medium-sized firms, it was more difficult to access senior-level executives in large, multi-national firms. In such cases, I tried to eliminate bias in the interviews by asking respondents to describe usual patterns of firm representatives. I also tried to phrase the interview questions in a way that encouraged respondents to speak from the perspective of the firm, rather than describe their own personal experiences or opinions. Taken together, these efforts should boost confidence regarding the representativeness of the study's findings with respect to patterns of firm-level search processes. In approaching potential respondents, I also tried to represent the structure and diversity of each of the four trade fairs. To this end, firms were randomly selected from the floor plan of the entire exhibition space for each trade fair, making sure that all of the different product or technology focus segments were covered.

Along with the interview responses, primary documents from trade fair exhibitors and participant observations were used to inform the coding of the variables; the quantitative data also offered some guidance on this front. The interview responses were used to create an original typology to classify different varieties of exhibitors' search processes (discussed in Chap. 5). This process involved careful examination of the interview transcripts in order to identify and make sense of different patterns in the search processes of firms. The different types of search behaviour relate to two different dimensions: first, the goal of the search process, and, second, the approach underlying the search process. The typology identifies four different categories of technological search behaviour: (i) unclear and unstructured; (ii) ambiguous yet optimistic; (iii) confident but unclear; and (iv) clear and structured. This typology, which ranges from completely unstructured, on the one extreme, to highly structured, on the other, does not claim to capture all of the nuances and complexities associated with technological search processes at trade fairs. Yet, it serves as a useful tool in distinguishing among main categories of search behaviours and related knowledge processes. Chapter 5 develops this much more fully when the findings from the qualitative analysis are discussed.

1.2.3.2 The Quantitative Study

For the quantitative analysis, I followed two different data collection strategies. The first involved distributing the questionnaire online to lists of exhibitors that had previously exhibited at one of the four selected trade fairs – *bauma, Electronica, EMO Hannover*, or the *International CES*. The second involved distributing the same questionnaire in-person during two of these same events. While the original plan was to rely entirely on data collected from the online survey, the initial response rate was very low, about 2%, which prompted a change in strategy. Thus, instead of relying exclusively on email communication, the same questionnaire was distributed in person to exhibitors at two of the previously selected trade fairs, held in subsequent years – EMO Hannover, 2011, and the International CES, 2012. These two events fit with the general rationale underlying the selection of the trade fairs; yet, due to temporal and financial constraints, I was only able to include two of the four originally selected trade fairs for the in-person data collection. The response rate from the in-person data collection procedure at EMO and CES was considerably higher, at 60% and 74%, respectively. After pooling the responses from both the online and in-person surveys, the total number of cases was 451. Of this total number, 105 represent U.S. firms and 60 represent German firms. Given that my interest in the quantitative analysis is to compare these two capitalist varieties, the present study focuses on these 165 cases.

I conducted statistical tests to determine if there were any significant differences between those that filled out the online versus the paper survey, by firm age, size, sector, and country. The findings revealed that German firms were slightly more likely than U.S. firms to fill out the paper survey – significant at a more modest 10% threshold. Further, with regards to sector, machinery firms demonstrated a greater likelihood than electronics firms to complete the paper survey ($p = 1\%$). While there may be different explanations for the above issues, one might involve the contextual differences associated with completing the survey by oneself in front of a computer versus filling it out while in the midst of the trade fair setting. This might have a greater impact on firms in the machinery sector. Regardless of the reasons, the above differences are a reality of the data and must be considered when conducting the analysis.

The self-administered questionnaire, which was used for both the online and in-person survey modes, contains a number of questions designed to evaluate the nature of firms' search processes at trade fairs. The first section of the questionnaire asks respondents to rank order the competencies they consider most valuable to the operations of the firm. It also asks them to rank the types of competencies they look for in potential partners at trade fairs. These competencies are: (*i*) *innovation focus,*

(ii) low-cost focus, (iii) top-of-the-line production, (iv) strong R&D capabilities, (v) marketing excellence, (vi) customized production, and (vii) continuously changing product line. A comparison of the two groups of competencies enabled a better understanding of the types of competencies firms search for and whether these competencies are generally similar to or different from their own. Through this, I was able to gain some insight into the nature of inter-firm search at trade fairs and the adoption of different pathways of technological development – i.e., toward specialization or standardization.

A second set of questions focused specifically on the character of inter-firm knowledge flows at trade fairs. These questions probe different aspects of knowledge flows, including where the firm gets knowledge, the type of knowledge it gets, to whom it provides knowledge, and the type of knowledge it provides. For each question, respondents were provided with six different possible responses[12] and asked to rank order them from least to most important. For this, I used a 7-point Likert scale, where 1 = "not important" and 7 = "most important". The logic behind the four knowledge questions is two-fold. First, such questions can help determine the character and importance of inter-firm knowledge interactions at trade fairs and, second, they can help elucidate the direction or pathways of such knowledge flows during trade fairs. Beyond these knowledge questions, the survey participants were also asked to rank order different activities in which their firm participates in order to learn about the industry/market. The list includes 11 activities, some of which are based on F2F communication (e.g., participation in trade fairs, lunch meetings with other firms), while others involve the use of online social media platforms (e.g., blogs, Facebook, Twitter).

The final section of the questionnaire was designed to obtain information on a wide range of firm characteristics, including age, size (number of employees), annual sales, country of origin, and the international/domestic focus of the firm's operations (See Appendix A for a copy of the questionnaire and Appendix C for descriptive results).

1.3 Research Agenda and Implications

The main agenda of this research is to demonstrate that international trade fairs serve to connect distinct political economies and that the exchanges of firms at these events largely reflect the institutional and political settings in which they operate. Such findings contribute to ongoing discussions in comparative political economy

[12] For the questions about where firms get knowledge and to whom they provide knowledge, the following six responses were provided: (i) existing customers; (ii) potential customers; (iii) competitors; (iv) existing suppliers; (v) potential suppliers; and (vi) other partners in research/distribution. For the questions about what type of knowledge is received or provided, the following responses were listed: (i) product-related; (ii) technology-related; (iii) market-related; (iv) industry-related; and (v) management-related. See Appendix A for the complete questionnaire.

by deepening our understanding of the potential future pathways of national political economies, as well as our understanding of the competitive pressures towards convergence on a single national model and the nature of firms' resilience in the face of such pressures.

As noted above, the present study is explicitly multi-disciplinary in nature. Central to adopting an interdisciplinary focus is a desire to encourage greater dialogue between the literatures on comparative capitalist varieties and economic geography, specifically, but also among studies in economics, organizational behaviour, and management. As Peck and Theodore (2007) observe, economic geographers and VoC scholars share common ground on a number of important fronts – theoretical, empirical, and methodological.[13] Yet, there have been remarkably few meaningful exchanges between the fields. For example, VoC scholars seldom acknowledge the contributions of economic geographers (see, for instance, Coates, 2005; Crouch, 2005), and the VoC approach is rarely discussed by scholars working in the field of economic geography (for important exceptions, however, see Christopherson, 2002; Bathelt & Gertler, 2005; Clark & Wojcik, 2007).

Beyond a desire to start exploring the possibilities for fruitful exchange between these fields, the insights generated from this study may have important implications for governments and policy-makers at different levels. As noted in the opening to this chapter, there is a clear need for countries around the world to develop new avenues for economic growth and competitiveness in the face of recent developments in the global political economy and the associated uncertainties that accompany such changes. In exploring the micro-foundations of inter-firm interactions and search during international trade fairs, this research adds new empirical evidence on the potential importance of these events in the global knowledge-based economy. Through this, the study also aims to articulate clear ways in which governments at different levels might support firms in developing, for instance, specific competencies related to innovation, as well as more effectively using trade fairs to fuel economic growth in their local, regional, and national state contexts.

The study also raises a number of important and related questions that could form the basis of an interesting and important future research agenda. How, for example, might policies best be designed to lower or reduce barriers to access such valuable distant knowledge sources? What is the role of governments (at local, regional, and national levels) in supporting firms' search processes through, for example, financial and/or organizational assistance to participate in international trade fairs? And, should governments be investing more resources into hosting competitive trade fairs and building attractive venues for such events, as has been done

[13] As Peck and Theodore (2007, 732–733) observe, VoC scholars and economic geographers typically share: "a substantial degree of skepticism concerning 'hyperglobalization' rhetoric and the associated embrace of convergence theses; a commitment to theoretically informed concrete research on distinctive, 'local' forms of contemporary capitalism and economic restructuring; a recognition of the institutionally mediated, socially embedded nature of economic structures and relations; and a pluralistic intellectual culture, marked by various degrees of divergence from, or dissonance with, neoclassical economics."

in places such as Las Vegas, Nevada, in the U.S.? How might international trade fairs be positioned as part of a policy package to attract key business actors and "buzz" to a city or city-region?

A greater focus on international trade fairs promises substantial contributions to the field of political science. In the realm of comparative political economy, for instance, there remain unanswered questions regarding the persistence of different national models of capitalism and how some political economies have shifted in incremental yet meaningful ways over time (e.g., Hall & Soskice, 2001; Crouch, 2005; Hall & Thelen, 2009; Mahoney & Thelen, 2010). This study builds on recent theoretical developments focused on the dynamic adjustments occurring within political-economic systems, many of which are still in the early stages of conceptual development and empirical testing (Streeck & Thelen, 2005; Hall & Thelen, 2009; Mahoney & Thelen, 2010). In this context, the present study aims to provide some empirically-grounded insights into how the search behaviours and interaction processes of firms at trade fairs might influence economic specialization patterns in relation to the application and production of technologies. Additionally, the study of international trade fairs could be particularly valuable for political scientists focused on processes of economic development in less-developed or transitioning political economies. A recent study on trade fair ecologies in the Asia-Pacific region (Bathelt & Zeng, 2014) points to the importance of international trade fairs in positioning countries within the realm of global competition and in the development of approaches and policies that might support their economic growth. These are but two examples of research agendas that could be enriched by a focus on trade fairs.

1.4 The Structure of the Argument

The next three chapters of this book lay out the theoretical foundations of the study. Following recent work in the Comparative Capitalisms (CC) tradition (Deeg & Jackson, 2007; Jackson & Deeg, 2008), a framework is proposed that aims to incorporate processes operating at the macro-, meso-, and micro-levels of analysis in an effort to develop a deeper understanding of the dynamic and often emergent processes by which institutional change can occur. Starting at the meso-level, Chap. 2 explores the existing and still quite fragmented literature on trade fairs. Paying particular attention to studies that emphasize the relationship between international trade fairs and processes of knowledge creation and exchange in the global economy, Chap. 2 conceptualizes trade fairs as critical meso-level structures that link micro- and macro-levels by mediating the exchange of technological knowledge between firms from across different capitalist varieties. This chapter encourages a view of international trade fairs as pivotal platforms that enable inter-firm interactions and knowledge exchange, and argues that these events represent an important and largely overlooked subject of inquiry within the field of comparative political economy. This study represents a first step in addressing this oversight. It encourages others working within the field of comparative political economy, and political science more broadly, to consider international trade fairs as an important area of theoretical and empirical investigation. This chapter also describes the selection of trade fair sites used in this study along with some contextual background on the trade fairs.

Chapter 3 shifts the discussion to the macro-level and to debates regarding the convergence or divergence of distinct national political economies, specifically with respect to patterns of technological specialization. This chapter offers an overview of several prominent contributions to the study of comparative capitalisms, most notably Hall and Soskice's (2001) seminal work on Varieties of Capitalism. Important to this discussion is the longstanding problem of institutional change within the context of comparative political economy research. Providing a stylized review of the existing literature in this area, Chap. 3 ultimately argues that the study of international trade fairs in the context of global political-economic processes can contribute important insights to ongoing debates in political science regarding the nature of institutional change, particularly with respect to recent theoretical developments on more incremental forms of institutional change (Thelen, 2004; Streeck & Thelen, 2005; Mahoney & Thelen, 2010; van der Heijden & Kuhlman, 2017).

Having discussed the macro- and meso-levels, Chap. 4 turns to the micro-level where I introduce the concept of "organized anarchies" and the "garbage-can" model of organizational decision-making (Cohen et al., 1972). The garbage-can model is useful not only in characterizing the search processes of firms at international trade fairs, but also in developing an empirically grounded examination of how specific micro-level processes (i.e., inter-firm interaction during trade fairs) might relate to and possibly shape other macro-level processes (i.e., national patterns of technological specialization) through mechanisms operating at the meso-level of the trade fair. Chapter 4 also discusses relevant theoretical findings in the literature on firms' technological search behaviours and in the recent work on behavioural economics.

Chapters 5 and 6 represent the empirical core of the analysis. Chapter 5 explores, in an in-depth, qualitative way, the technological search and interaction processes of firms at four leading North American trade fairs. Drawing on the garbage-can model of organizational choice (Cohen et al., 1972), as well as key ideas from the literature on organizational search and the subfield of behavioural economics, Chap. 5 examines the nature and dynamics of these processes and aims to link the findings to broader patterns of technological specialization at the level of the national state. In this chapter, I present an original typology of firms' search behaviours at trade fairs and use this typology to compare the different types of search observed during these events. This is helpful in developing a deeper understanding of the specific trajectories and potential implications of inter-firm knowledge flows at international trade fairs.

Chapter 6 relies largely on quantitative methods and aims to address fundamental questions pertaining to the broad similarities and differences between German (CME) and U.S. (LME) firms exhibiting at international trade fairs. This chapter explores the activities used by firms from different countries for technological learning, the core competencies of the different firms, and the competencies such firms seek out in potential business partners. It develops and tests the hypothesis that German and U.S. firms will differ in significant ways in these areas. An important objective of Chap. 6 will be to provide a stronger basis of descriptive empirical work on the differences between these countries within the realm of inter-firm interactions, particularly knowledge flows. In developing such foundations, this chapter develops the argument that the interactions of firms during international trade fairs are largely guided by differences in the institutional arrangements and production contexts of the specific capitalist variety in which they are embedded. Extending the basis for the qualitative findings on search processes presented in Chap. 5, Chap. 6 suggests that inter-firm interactions at trade fairs support, rather than undermine, the continued divergence of national political economies in terms of their technological specialization patterns.

Finally, Chap. 7 concludes by considering the implications of a focus on international trade fairs for the field of political science. This includes implications for the comparative study of political economies as well as research focused on public policy and policy-making in the context of the globalizing knowledge-based economy. With respect to comparative political economy studies, for instance, international trade fairs offer an incredibly useful empirical setting in which to explore cross-national interaction patterns and dynamics using the firm as the unit of analysis. As this chapter discusses, trade fairs provide a unique window to ongoing questions and debates in different areas of political science. Given the unprecedented role of the COVID-19 pandemic on the global political economy, including restrictions on mass gatherings and international travel, Chap. 7 also reflects on how the pandemic has impacted the international trade fair industry and how it might shape the future organization, dynamics, and relevance of these events.

References

Armstrong, M., & Huck, S. (2010). *Behavioral economics as applied to firms: A primer*. CESifo Working Paper Series No. 2973.

Armstrong, M., & Huck, S. (2014). Behavioral economics and antitrust. In R. D. Blair & D. D. Sokol (Eds.), *The Oxford handbook of international antitrust economics* (Vol. 1, pp. 205–228). Oxford University Press.

Arundel, A., & Constantelou, A. (2006). Conventional and experimental indicators of knowledge flows. In Y. Caloghirou, A. Constantelou, & N. S. Vonortas (Eds.), *Knowledge flows in European industry* (pp. 45–66). Routledge.

Asheim, B. T. (1999). Interactive learning and localised knowledge in globalizing learning economies. *GeoJournal, 49*(4), 345–352.

AUMA. (2009). *Die Messewirtschaft: Bilanz 2008 (the trade show economy: 2008 balance)*. AUMA.

Bacchus, J. (2022, April 26). *Biden and trade at year one: The reign of polite protectionism*. Policy Analysis no. 926, Cato Institute. Retrieved on June 8, 2022, from https://www.cato.org/policy-analysis/biden-trade-year-one

Baschuk, B. (2022). *What 'friend-shoring' means for trade in a less-friendly world*. The Washington Post. Retrieved on June 8, 2022, from https://www.washingtonpost.com/business/what-friend-shoringmeans-for-trade-in-a-less-friendly-world/2022/06/22/5c3d4f30-f1e3-11ec-ac16-8fbf7194cd78_story.html

Bathelt, H., & Gertler, M. S. (2005). The German variety of capitalism: Forces and dynamics of evolutionary change. *Economic Geography, 81*, 1–9.

Bathelt, H., & Glückler, J. (2011). *The relational economy: Geographies of knowing and learning*. Oxford University Press.

Bathelt, H., & Schuldt, N. (2008). Between luminaires and meat grinders: International trade fairs as temporary clusters. *Regional Studies, 42*(6), 853–868.

Bathelt, H., & Zeng, G. (2014). The development of trade fair ecologies in China: Case studies from Chengdu and Shanghai. *Environment and Planning, 46*(3), 511–530.

Bathelt, H., Golfetto, F., & Rinallo, D. (2014). *Trade shows in the globalizing knowledge economy*. Oxford University Press.

Bienkowska, D., Lundmark, M., & Malmberg, A. (2011). Brain circulation and flexible adjustment: Labour mobility as a cluster advantage. *Geografiska Annaler: Series B, Human Geography, 93*(1), 21–39.

Borghini, S., Golfetto, F., & Rinallo, D. (2004). Using anthropological methods to study industrial marketing and purchasing: An exploration of professional trade shows. In *Presented at the industrial marketing purchasing conference*. Retrieved on May 11, 2022, from http://www.impgroup.org/paper_view.php?viewPaper=4505.

Bruff, I., Ebenau, M., & May, C. (2015). Fault and fracture? The impact of new directions in comparative capitalisms research on the wider field. In M. Ebenau, I. Bruff, & C. May (Eds.), *New directions in comparative capitalisms research: Critical and global perspectives* (pp. 28–44). Palgrave Macmillan.

Christopherson, S. (2002). Why do national labor market practices continue to diverge in the global economy? The 'missing link' of investment rules. *Economic Geography, 78*(1), 1–20.

Clark, G. L., & Wójcik, D. (2007). *The geography of finance: Corporate governance in a global marketplace*. Oxford University Press.

Coates, D. (2005). Paradigms of explanation. In D. Coates (Ed.), *Varieties of capitalism, varieties of approaches* (pp. 1–25). Palgrave Macmillan.

Cohen, M. D., March, J. G., & Olsen, J. P. (1972). A garbage can model of organizational choice. *Administrative Science Quarterly, 17*(1), 1–25.

Crouch, C. (2005). *Capitalist diversity and change: Recombinant governance and institutional entrepreneurs*. Oxford University Press.

Crouch, C., & Streeck, W. (1997). *Political economy of modern capitalism: Mapping convergence and diversity*. Sage.

Cyert, R. M., & March, J. G. (1963). *A behavioral theory of the firm*. Prentice-Hall.

Deeg, R., & Jackson, G. (2007). Towards a more dynamic theory of capitalist variety. *Socio-Economic Review, 5*, 149–179.

Ebenau, M., Bruff, I., & May, C. (2015). *New directions in comparative capitalisms research: Critical and global perspectives*. Palgrave Macmillan.

Flew, T. (2020). Globalization, neo-globalization and post-globalization: The challenge of populism and the return of the national. *Global Media and Communication, 16*(1), 19–39.

Frum, D. (2022, March 2). *This is no time for protectionism*. The Atlantic. Retrieved on June 8, 2022, from https://www.theatlantic.com/ideas/archive/2022/03/biden-economic-nationalism-state-of-the-union/623325/

Gavetti, G., & Levinthal, D. A. (2000). Looking forward and looking backward: Cognitive and experimental search. *Administrative Science Quarterly, 45*(1), 113–137.

Gertler, M. S. (2003). Tacit knowledge and the economic geography of context, or the undefinable tacitness of being (there). *Journal of Economic Geography, 3*(1), 75–99.

Gibson, R., & Bathelt, H. (2010). Understanding the dynamics of specialization and diffusion processes across capitalist varieties: A conceptual intervention regarding the role of international trade fairs. *SPACES* 2010–4. Retrieved on May 8, 2022, from http://www.spaces-online.com.

Golfetto, F., & Rinallo, D. (2013). Trade show events, from live to digital: Towards the development of new competencies. *SPACES, 12*, 119–154.

Golfetto, F., & Rinallo, D. (2014). The evolution of trade show systems: Lessons from Europe. In H. Bathelt & G. Zeng (Eds.), *Temporary knowledge ecologies: The rise and evolution of trade fairs in the Asia-Pacific region* (pp. 42–66). Edward Elgar.

Goodhart, D. (2017). *The road to somewhere: The populist revolt and the future of politics*. Hurst Publishers.

Grant, R. M., & Baden-Fuller, C. (2004). A knowledge accessing theory of strategic alliances. *Journal of Management Studies, 41*, 61–84.

Griliches, Z. (1992). Introduction. In Z. Griliches (Ed.), *Output measurement in the service sectors* (pp. 1–22). University of Chicago Press.

Hall, P. A., & Soskice, D. (2001). An introduction to varieties of capitalism. In P. A. Hall & D. Soskice (Eds.), *Varieties of capitalism: The institutional foundations of comparative advantage* (pp. 1–68). Oxford University Press.

Hall, P. A., & Thelen, K. (2009). Institutional change in varieties of capitalism. *Socio-Economic Review, 7*(1), 7–34.

Ignatius, A. (2017). The truth about globalization. *Harvard Business Review, 95*(4). Retrieved on May 8, 2022, from https://hbr.org/2017/07/the-truth-about-globalization.

Jackson, G., & Deeg, R. (2006). *How many varieties of capitalism? Comparing the comparative institutional analyses of capitalist diversity*. MPIfG Discussion Paper 06/2, Max Planck Institute for the Study of Societies.

Jackson, G., & Deeg, R. (2008). Comparing capitalisms: Understanding institutional diversity and its implications for international business. *Journal of International Business Studies, 39*(4), 540–561.

Jones, C. (2020, November 17). Was hyperglobalisation an anomaly? Maybe, but don't you dare say "slowbalisation". *Financial Times*.

Katila, R., & Ahuja, G. (2002). Something old, something new: A longitudinal study of search behavior and new product introduction. *Academy of Management Journal, 45*(8), 1183–1194.

Katzenstein, P. (2006). Denmark and small states. In J. L. Campbell, J. A. Hall, & O. K. Pedersen (Eds.), *National identity and the varieties of capitalism* (pp. 433–440). McGill-Queen's University Press.

Kitschelt, H., Lange, P., Marks, G., & Stephens, J. (Eds.). (1999). *Continuity and change in contemporary capitalism*. Cambridge University Press.

Lampel, J., & Meyer, A. D. (2008). Guest editors' introduction: Field-configuring events as structuring mechanisms: How conferences, ceremonies, and trade shows constitute new technologies, industries, and markets. *Journal of Management Studies, 45*(6), 1025–1035.

Lampel, J., & Shapira, Z. (1995). Progress and its discontents: Data scarcity and the limits of falsification in strategic management. In P. Shrivastva (Ed.), *Advances in strategic management* (Vol. 12, pp. 113–150).

Mahoney, J., & Thelen, K. (2010). A theory of gradual institutional change. In J. Mahoney & K. Thelen (Eds.), *Explaining institutional change: Ambiguity, agency, and power* (pp. 1–37). Cambridge University Press.

Maskell, P., & Malmberg, A. (1999a). The competitiveness of firms and regions: 'Ubiquitification' and the importance of localised learning. *European Urban and Regional Studies, 6*(1), 9–25.

Maskell, P., & Malmberg, A. (1999b). Localised learning and industrial competitiveness. *Cambridge Journal of Economics, 23*(2), 167–185.

Maskell, P., Bathelt, H., & Malmberg, A. (2006). Building global knowledge pipelines: The role of temporary clusters. *European Planning Studies, 14*(8), 997–1013.

Meyer, B. (2022). *A playbook against populism? Populist leadership decline in 2021*. Tony Blair Institute for Global Change. Retrieved on June 7, 2022, from https://institute.global/sites/default/files/articles/A-Playbook-Against-Populism-Populist-Leadership-in-Decline-in-2021.pdf

Micek, G. (2016). Shortcomings and weaknesses in understanding and measuring knowledge interactions. In G. Micek (ed.), *Understanding innovation in emerging economic spaces: Global and local actors, networks and embeddedness* (pp. 58–70), Routledge.

Nonaka, I. (1994). A dynamic theory of organizational knowledge creation. *Organization Science, 5*(1), 14–37.

Nonaka, I., Toyama, R., & Konno, N. (2000). SECI, *Ba* and leadership: A unified model of dynamic knowledge creation. *Long Range Planning, 33*(1), 5–34.

Peck, J., & Theodore, N. (2007). Variegated capitalism. *Progress in Human Geography, 31*(6), 731–772.

Peri, G. (2005). Determinants of knowledge flows and their effect on innovation. *Review of Economics and Statistics, 87*(2), 308–322.

Rodrik, D. (2011). *The globalization paradox: Democracy and the future of the world economy*. W. W. Norton.

Rosenkopf, L., & Nerkar, A. (2001). Beyond local research: Boundary-spanning, exploration, and impact in the optical disk industry. *Strategic Management Journal, 22*(4), 287–306.

Sotarauta, M., Ramstedt-Sen, T., Seppänen, S. K., & Kosonen, K.-J. (2011). Local or digital buzz, global or national pipelines: Patterns of knowledge sourcing in intelligent machinery and digital content services in Finland. *European Planning Studies, 19*(7), 1305–1330.

Storper, M. (1997). *The regional world: Territorial development in a global economy*. Guilford Press.

Streeck, W., & Thelen, K. (2005). Introduction: Institutional change in advanced capitalist economies. In W. Streeck & K. Thelen (Eds.), *Beyond continuity: Institutional change in advanced capitalist economies* (pp. 1–39). Oxford University Press.

Streeck, W., & Yamamura, K. (2003). Introduction: Convergence or diversity? In K. Yamamura & W. Streeck (Eds.), *The end of diversity? Prospects for German and Japanese capitalism* (pp. 1–50). Cornell University Press.

Subramanian, A., & Kessler, M. (2013). *The hyperglobalization of trade and its future*. Working Paper Series (WP 13-6). Peterson Institute for International Economics. Retrieved on May 11, 2022, from https://www.piie.com/publications/wp/wp13-6.pdf

The Economist. (2019a). *The steam has gone out of globalisation*. Retrieved on May 11, 2022, from https://www.economist.com/leaders/2019/01/24/the-steam-has-gone-out-of-globalisation

The Economist. (2019b). *Globalisation has faltered*. Retrieved on May 11, 2022, from https://www.economist.com/briefing/2019/01/24/globalisation-has-faltered

The Economist. (2022). *The tricky restructuring of global supply chains*. Retrieved on June 18, 2022, from https://www.economist.com/leaders/2022/06/16/the-tricky-restructuring-of-global-supply-chains

Thelen, K. (2003). How institutions evolve: Insights from comparative-historical analysis. In J. Mahoney & D. Rueschemeyer (Eds.), *Comparative historical analysis in the social sciences* (pp. 208–240). Cambridge University Press.

Thelen, K. (2004). *How institutions evolve: The political economy of skills in Germany, Britain, the United States, and Japan*. Cambridge University Press.

van Bergeijk, P. A. G. (2019). *Deglobalisation 2.0: Trade and openness during the great depression and the great recession*. Edward Elgar.

van der Heijden, J., & Kuhlmann, J. (2017). Studying incremental institutional change: A systematic and critical meta-review of the literature from 2005 to 2015. *Policy Studies Journal, 45*(3), 535–554.

Visser, E.-J. (2009). The complementary dynamic effects of clusters and networks. *Industry and Innovation, 16*(2), 167–195.

Weterings, A., & Ponds, R. (2009). Do regional and non-regional knowledge flows differ? An empirical study on clustered firms in the Dutch life sciences and computing services industry. *Industry and Innovation, 16*(1), 11–31.

Winter, S. G. (1984). Schumpeterian competition in alternative technological regimes. *Journal of Economic Behavior & Organization, 5*(3–4), 287–320.

Yamamura, K., & Streeck, W. (Eds.). (2003). *The end of diversity? Prospects for German and Japanese capitalism*. Cornell University Press.

Chapter 2
International Trade Fairs: Mediators of Institutional Change

2.1 Introduction

Why should we focus our analytical lens on international trade fairs? What is their significance in the context of the global political economy and in relation to theories of institutional change? What sets them apart from other critical sources of economic dynamism? This chapter aims to address these questions in two parts. The first part of the chapter sets the stage by describing the basic anatomy of the contemporary trade fair as we know it today. Having established a clearer sense of what these events are and what they do, I then turn to the treatment of trade fairs in the academic literature. From a practical standpoint, this literature review aims to highlight important changes in the function and character of trade fairs over time, while also pointing to the historical significance of these events in the development of, for instance, early systems of commercial law.[1] From a theoretical perspective, the literature review draws particular attention to recent novel conceptualizations of trade fairs in fields such as economic geography, while also noting the relative absence of serious research on trade fairs in other areas of study – notably, political science. Against this backdrop, I argue that trade fairs not only serve as an increasingly vital source of economic dynamism in the global knowledge-based political economy, but that they can also act as important mediators of incremental institutional change, specifically with respect to national patterns of technological specialization/standardization. In developing these hypotheses, I hope to foster greater dialogue between the fields of political economy and economic geography and invite students and practitioners interested in the dynamics of institutional change to consider trade fairs as important sites of theoretical and empirical relevance.

---

[1] For a more comprehensive description of the origins and evolution of trade fairs, see Bathelt et al. (2014, 17–39).

R. Gibson, *International Trade Fairs and Inter-Firm Knowledge Flows*,
https://doi.org/10.1007/978-3-031-20557-6_2

In the second part of the chapter, I situate the discussion of trade fairs within the context of other commonly cited sources of economic dynamism. This literature consists of several partial theories that identify and aim to understand the role of different factors in shaping economic progress. Through a brief review of some of the main determinants of economic dynamism, I highlight the concept of "knowledge-based" or "knowledge-driven" economic dynamism, developed in the work of Arvanitidis and Petrakos (2011). Using this conceptualization as a starting point, I suggest that international trade fairs represent a unique and still largely under-studied source of economic dynamism. More specifically, the relational and knowledge creation and exchange attributes of trade fairs produce an environment that helps to structure, strengthen, and promote multiple determinants of economic dynamism (e.g., innovation capacity, human capital) simultaneously. In this way, I suggest that leading international trade fairs may take on an increasingly critical role in driving economic dynamism in the context of the knowledge-based global political economy.

2.2 Trade Fairs in Theoretical Perspective

Trade fairs have been described as organized "*large-scale professional gatherings*" that are "bound by place and time, where a large number of professionals within defined industries assemble to network and to exchange both information and goods" (Nyqvist et al., 2017, 2 [emphasis original]). They are temporary marketplaces that bring together different groups of suppliers (or exhibitors) from a particular industry or value chain to present and discuss their own products and services as well as observing those of others (Black, 1986; Rinallo & Golfetto, 2011). Trade fairs are typically held anywhere from one to several days, and follow a specific rhythm or temporal frequency – e.g., annual, biennial, triennial – in response to the industry's particular innovation cycle (Bathelt et al., 2014).[2]

2.2.1 The Basic Anatomy and Functions of Trade Fairs

The work of Bathelt et al. (2014) identifies four key characteristics that distinguish the modern-day trade fair from other types of events, such as conferences or corporate events. Trade fairs are, first and foremost, *commercial events* that target

[2]Trade fairs for the fashion or textile industries, for example, usually follow a bi-annual trade fair rhythm, whereas machine tool trade fairs generally occur less frequently – once every 2, 3, or 4 years (Bathelt et al., 2014).

professional audiences, as opposed to the general public or consumers.[3] Although many trade fairs include entertainment or educational elements, these elements are secondary to the commercial aspect of the event. It is perhaps this emphasis on the commercial that has discouraged scholars in fields like political science, for example, from taking these events more seriously as an area of relevant empirical inquiry. In addition to being commercial events that focus on professional audiences, trade fairs also embody a form of *collective promotion* in the sense that they bring together a large number of firms from the same or related industry or technology field. Finally, the activities and interactions of firms at trade fairs revolve around material or *physical artefacts*, such as product samples and demonstrations (Bathelt et al., 2014). The centrality of such physical artefacts is critical to the creation of a specific "information and communication ecology" that is distinctive to the trade fair setting, known as "global buzz" (Bathelt & Schuldt, 2010). Global buzz, which will be discussed in more detail later on in this chapter, contributes to an environment that is characterized by rich and varied sources of knowledge that can be obtained and shared through multiple avenues simultaneously. Together, the above four features – *commercial events, professional audiences, collective promotion,* and *physical artefacts* – define the basic contours of the modern-day trade fair.

Learning and knowledge exchange are important motivations for participating in international trade fairs. The industry- or technology-specific focus of these events provides a unique setting where economic agents working within the same professional community come together for the purpose of exchanging knowledge about existing and future developments in their industry or technological field (Rinallo et al., 2007; Bathelt & Gibson, 2015). Belonging to the same professional community can have considerable advantages when it comes to enabling processes of learning and knowledge exchange; members of the same professional community typically have similar knowledge bases, speak the same industry-based "language", and identify with a shared industry "culture" or value system. In the trade fair context, face-to-face (F2F) communication with other participants (for example, customers, competitors, and potential business partners) along with close observations of, or interactions with, material artefacts are the primary means through which learning and knowledge exchange take place (Bathelt & Schuldt, 2010; Bathelt et al., 2014; Bathelt & Gibson, 2015).

[3] It is important to note the difference between business-to-business (B2B) fairs and business-to-consumer (B2C) fairs. The difference relates to the specific audiences targeted by the trade fair. B2B fairs bring together economic agents from different stages of the value chain for the purpose of presenting, scanning, and discussing new trends or technological developments in the industry. Beyond their marketing and sales functions, B2B fairs play a key role in the creation and exchange of knowledge related to, for instance, the development of new products, services, or technologies, as well as changes in the institutional environment, and industry or market trends (Bathelt et al., 2014). B2C fairs, on the other hand, target consumers. Participants at B2C fairs are sometimes described as being less interested in learning opportunities than B2B events, and more focused on simply having an enjoyable experience. Because the value-chain character of these fairs is sometimes lost or diminished, B2C events have been deemed less critical to the development of the industry or technology field than B2B events (Bathelt et al., 2014).

In the context of learning and knowledge exchange, the importance of trade fairs as an area of intellectual inquiry becomes relevant to a wide number of fields or disciplines including, and of particular interest here, political science and political economy. As Lundvall and Johnson (2016, 107) point out, if we accept that "knowledge *is* the most fundamental resource in our contemporary economy and that learning is therefore the most important process", this has far-reaching implications for political-economic theory as well as the institutional set-up of contemporary capitalist economies. While Lundvall and Johnson (2016) do not refer directly to trade fairs in addressing the traditional "market versus state intervention" debate from a "learning economy" perspective (Lundvall & Johnson, 1994), in this chapter, I argue that, from both a conceptual and empirical standpoint, trade fairs provide critical insights into the precise mechanisms by which firms perform or engage in learning processes. This argument will be developed not in relation to discussions regarding planned versus market economies, but in relation to debates regarding the convergence/divergence of distinct capitalist varieties as well as broader questions regarding the nature and dynamics of institutional change.

Whereas the above discussion stresses the significance of trade fairs in the global knowledge-based economy, it is important to note that these events are far from new. Indeed, trade fairs have been around in one form or another since at least the medieval period, and perhaps even well before that (Braudel, 1992; Moeran, 2011; Bathelt et al., 2014).[4] Yet, the analysis of these events in the academic literature is remarkably fragmented across different academic disciplines. The discussion below aims to provide an overview of the treatment of trade fairs in different fields of study, with particular attention being paid to recent innovative approaches to the conceptualization and empirical study of these events in economic geography and, more recently, marketing studies. This review also notes how other disciplines, specifically political science, have overlooked trade fairs as a valuable site for empirical research. In doing so, I call for increased attention to these events in the field of political science, in general, and the study of comparative political economy, in particular. More specifically, I aim to demonstrate that the topic of trade fairs should be of considerable interest to scholars concerned with the dynamics of the global political economy and the impact of such dynamics on a wide range of political, economic, and social phenomena.

[4] Braudel (1992) traces the trade fairs of Lyon, for instance, back to 172 A.D., and the fairs of Troyes to Roman times. Bathelt et al. (2014) trace the origin of European trade fairs as far back as the era of the Celtic pilgrimages, before the founding of Rome. Moreover, trade fairs in India and China have been traced to 4000 BC and 3000 BC, respectively (Allix, 1922; Bathelt et al., 2014).

2.2.2 *Trade Fairs in Historical Studies*

Historians have long been interested in the role of trade fairs as key places for economic and cultural exchange. This literature offers detailed historical analyses of specific significant events, including the fairs of Champagne, Flanders, and Lyon, in France, the Hamburg, Frankfurt, and Leipzig trade fairs in Germany, as well as the trade fairs of the English Midlands (Face, 1958; Verlinden, 1963; Bautier, 1970; Munro, 2001). This literature tells us that the fairs of the medieval period, such as the Champagne fairs (Face, 1958) played a key role in what was then already an international economy (Moeran, 2011). During this period, the development of trade fairs occurred at significant geographic points, such as the intersection of trade routes, for example, Leipzig (Flood, 2007; Moeran, 2011). In the context of medieval Europe, historians have documented the origins of trade fairs dating back to ancient civilizations and provided valuable descriptions of how the power of trade networks, such as the *Medici*, *Fugger* and *Hanse*, were built around trade fairs. This work identifies the early "commodity fairs" as places where traders from other regions gathered to sell their products to individual local customers during particular periods (such as Christian holidays) in response to protectionist trade policies (Allix, 1922).

Historians have remarked upon the international character of even the earliest types of trade fairs. During the Middle Ages, for instance, specific security measures were developed for merchants travelling to international fairs, such as a unique passport that guaranteed both merchants and their goods unrestricted passage on defined routes at specified times (Allix, 1922). Similar security assurances were offered on the actual fairgrounds. Under the so-called "peace of the fair", commercial transactions between merchants from different – sometimes competing or hostile – countries were protected by kings and feudal lords (Allix, 1922). The "peace of the fair" model has been thought to influence early systems of commercial law and international agreements (Milgrom et al., 1990). Beyond the issue of security, the international character of these early fairs was visible in the extent to which different national currencies were circulated at the large fairs: "there was a money exchange and even rudimentary forms of money-clearing" (Allix, 1922, 540). Attracting merchants from many different countries was deemed essential to the success and longevity of these early fairs. During this period in history, close geographical proximity to the borders of other countries was central to obtaining a strong international presence at trade fairs. As historical analyses demonstrate, the success of the trade fairs of Lyons, for instance, can be largely attributed to their proximity to the Swiss and Italian borders (Allix, 1922).

The first modern trade fair began in Leipzig in 1894. This marked a major turning point in the organizational model of trade fairs, as these events shifted away from the earlier fairs based on commodities and towards the contemporary "sample fairs" (Carreras & Torra, 2005). Unlike the traditional commodity fairs, the new sample fairs became events where only samples of products were displayed. At these sample fairs, customers placed orders for products, rather than having goods distributed to them directly at the fair. This contemporary form of trade fair was first introduced by Leipzig's *Mustermesse* – a fair in which exhibitors presented and discussed only samples of their commodities, while the products themselves were later shipped to the customer's home base (Fischer, 1992; Rodekamp, 2003). While Leipzig is recognized as the "birthplace" of the sample fair, it did not take long before the fairs in Lyon, London, Paris, and eventually all other fairs, followed suit (Bathelt et al., 2014). This fundamental change in the character of trade fairs was largely prompted by the development of new modes of transport and new telecommunications devices, which challenged traditional methods of organizing trade (Carreras & Torra, 2005). The relaxation of trade restrictions required new sets of institutions to be built in order to offer greater security in financial transactions and contractual guarantees. From this, trade fair rules eventually became an important foundation for the development of European business law (Maskell et al., 2006).

As the above review of the historical literature suggests, historians have provided rich documentation of the origins and evolution of trade fairs. Some of these studies have also noted the relationship of trade fairs to other important historical developments in, for example, systems of commercial law (Milgrom et al., 1990). This research enhances our understanding of the influence of trade fairs from an historical perspective and provides a solid foundation upon which other disciplines, such as political science or political economy, could build.

2.2.3 Trade Fairs in Marketing and Business Administration

Trade fairs have been an important area of research in the marketing and business administration literatures since the 1980s. Until very recently, however, such studies have adopted a fairly narrow view of these events as places for selling products, generating business leads, and negotiating contracts. Indeed, much of the mainstream marketing literature is preoccupied with evaluating the benefits of participating in trade fairs and developing clear performance indicators (Kerin & Cron, 1987; Sashi & Perretty, 1992; Gopalakrishna & Lilien, 1995; Dekimpe et al., 1997; Hansen, 2004). While most of these studies emphasize sales or qualified leads as the primary goal of trade fair participation, some scholars suggest that trade fairs also serve important non-sales objectives, such as gathering information about the market, monitoring competitors, and identifying prospective distributors or buyers (Bonoma, 1983; Kerin & Cron, 1987; Shoham, 1999; Seringhaus & Rosson, 2004). Another related focus of the marketing literature involves measuring the different activities of exhibitors at different phases – pre-fair, at-the-fair, and post-fair (Dekimpe et al., 1997; Weisgal, 1998; Seringhaus & Rosson, 2004). This literature reveals a highly fragmented approach to the analysis of such exhibitor activities and often fails to develop clear linkages among the various activities, phases, and performance outcomes (see, however, Seringhaus & Rosson, 2004).

A few empirical studies have contributed to our knowledge of trade fairs in areas such as the role of trade fairs in export promotion (Bello & Barksdale, 1986) and the nature of trade fair buyer behaviour (Bello, 1992). Yet, for the most part, the study of trade fairs in marketing and business administration consists of largely normative and prescriptive pieces, which suggest that trade fairs are valuable promotional tools when appropriately managed (Bonoma, 1983; Belizzi & Lipps, 1984; Sashi & Perretty, 1992; Shoham, 1992; Reid & Plank, 2004). Despite its prescriptive and normative orientation, this body of research clearly demonstrates that trade fairs are heavily relied upon across industries and by firms of all sizes (Reid & Plank, 2004).

One might argue that the prescriptive orientation of this literature has discouraged researchers in political science and other fields from recognizing or investigating the wider macroeconomic implications of international trade fairs. A similar argument might be made with respect to the general characterization of these events as focused mainly on commercial or managerial aspects of selling products. This study aims to demonstrate that international trade fairs are not simply events where firms come to promote and compare their products; they are also settings where political-economic systems, institutional practices and conditions, as well as changes in the political landscape are openly laid out. These aspects are significant and a key part of the discussions and interactions of trade fair participants.

Very recently, a small number of studies in marketing have begun to move away from the conventional themes and approaches discussed above. The work of Rinallo and Golfetto (2011), for example, stands out insofar as it aims to advance our understanding of the specific learning and interaction practices employed by specific actors, such as professional and entrepreneurial associations, at European Union clothing fabric trade fairs. In contrast to the prescriptive, "how-to", nature of much of the marketing literature, this research contributes new insights regarding the spatial implications of innovation and interaction processes. Moreover, one of the most comprehensive recent books on the subject of trade fairs involves a collaboration between scholars in the fields of marketing and economic geography (Bathelt et al., 2014). This work promotes an original view of trade fairs as rich sites of knowledge and learning that facilitate globalization processes, rather than merely being the product of such processes. This work is a testament not only to the potential importance of trade fairs in the global political economy, but also to the utility of adopting an interdisciplinary perspective to address new and ongoing academic questions.

2.2.4 Trade Fairs in Organization and Management Studies

In recent years, studies in organization and management have been interested in the nature and dynamics of "field-configuring events" (FCEs) – that is, "temporary organizations such as trade shows, professional gatherings, technology contests, and business ceremonies that encapsulate and shape the development of professions, technologies, markets, and industries" (Lampel & Meyer, 2008, 1026; see also Meyer et al., 2005).[5] FCEs are defined as sites in which diverse actors with diverse goals come together periodically, or sometimes only once, to present new products, discuss and advance industry standards, build social networks, exchange information, and conduct business (Lampel & Meyer, 2008). Although trade fairs are listed among the representative examples of such events (Lampel & Meyer, 2008), there have been few empirical studies applying the concept of FCEs specifically to the trade fair setting – a notable exception to this is Moeran and Pedersen's (2011) work on cultural fairs.

On the face of it, many of the core characteristics of FCEs[6] closely resemble the trade fair environment. Yet, upon closer examination, it seems that most trade fairs are unlikely to produce field configuration in the sense of an abrupt or discontinuous shift in an industry's technological or consumption pattern (Gibson & Bathelt, 2014). This is due to the decentralized character of knowledge flows that occur at trade fairs. The decentralized nature of knowledge flows at these events tends to produce a more nuanced impact on the relevant field, resulting in a process that may be better described as field-reproducing, rather than field-configuring (Gibson & Bathelt, 2014). That being said, recent research in organizational behaviour and management studies clearly demonstrates an increasing interest in the subject of trade fairs and recognition of their potentially transformative effects. Moreover, the literature on FCEs has an important connection to political science, as it investigates the conditions under which new fields emerge, as well as how institutional and political conditions are established and negotiated. Recent work has characterized UN conferences, for example, as field-configuring events, which can serve as "catalysts of change, especially as organizations and governments struggle to develop global solutions to complex problems" (Hardy & Maguire, 2010, 1365). In the growing FCEs literature, some studies have looked specifically at UN climate conferences, investigating the nature of institutional change – or lack thereof – in transnational climate policy (Schüssler et al., 2014). Other research in this literature has examined a key 1944 "Congress of Jewish Lawyers" in pre-state Israel, suggesting that this particular field-configuring event was instrumental in shaping the Israeli justice system (Oliver & Montgomery, 2008). Despite the clear political/policy relevance of this research, the concept of field-configuring events has not yet been picked up in the political science literature.

[5] This term was coined by Lampel and Meyer (2008) to refer to organized events such as trade fairs, professional conferences, festivals, governmental hearings, and public business ceremonies. The basic assumption is that interactions in temporally and spatially bounded contexts such as these enable both planned and unplanned processes that may have a profound influence on the future development of the field in which these events are situated.

[6] The key characteristics of FCEs may be outlined as follows: (i) they bring together in a single location geographically and organizationally diverse actors; (ii) their duration is limited, spanning a few hours to at most a few days; (iii) they offer unstructured opportunities for face-to-face interaction among participants; (iv) they feature and heavily depend upon ceremonial or dramaturgical activities; (v) they represent occasions for information exchange and collective sense-making; and (vi) they help to generate social and reputational resources that can later be used elsewhere (Lampel & Meyer, 2008).

2.2.5 *Trade Fairs in Economic Geography*

In the mainstream economic geography literature, research on trade fairs has largely focused on the local economic impact of these events through the use of regionalized input-output models that compute multiplier effects, or on the use of trade fairs to generate a local supplier sector across a variety of support industries and services. Generally speaking, the field of economic geography has been largely preoccupied with material indicators of economic activity, as seen in the vast literature on urban agglomerations, industrial districts, clusters, as well as innovation systems at different geographic levels (see, for example, Cooke & Morgan, 1998; Porter, 1990; Camagni, 1991). While several studies have focused on the conditions underlying agglomeration, much of the mainstream literature has failed to adequately investigate the role of knowledge flows or the nature and extent of the linkages between specific permanent entities, such as clusters (Bathelt et al., 2014). Very recently, however, we have seen a shift in scholarly focus "from structures to flows, from material characteristics to knowledge, and from permanent to temporary settings" (Bathelt et al., 2014, 267). In the context of this intellectual shift, scholars, particularly those working on the topic of "temporary clusters" and "temporary markets", have begun to view trade fairs as valuable sites of knowledge creation and exchange (Bathelt et al., 2004; Borghini et al., 2004; Maskell et al., 2006). From this perspective, international trade fairs represent crucial platforms for networking, learning, and creating and exchanging knowledge in the global knowledge-based economy. This is because trade fairs facilitate temporary, yet very concentrated and intense interactions among economic agents from all over the world, thereby breaking down traditional geographic boundaries and their associated challenges.

According to this body of work, ongoing F2F meetings with other trade fair participants creates opportunities for firms to establish or strengthen relationships with business partners, customers, suppliers, and even competitors. Repeat participation at the same event, for example, supports the development of trust relations and assists with the spread of reputational and behavioural cues among firms. This can lead to greater transparency and openness in business interactions, as well as the development of shared experiences and collective codes of conduct. The trade fair setting also allows firms to systematically obtain information and knowledge from these various agents, including details about their technological and strategic choices (Bathelt & Schuldt, 2008). There are multiple ways in which firms can acquire knowledge during these events. Information may be gathered through direct interaction with other firms at their exhibit stands. It may also be revealed through third parties, when a firm's customer, for example, leaks technological knowledge about a specific competitor. Considerable knowledge may also be obtained through careful observation and comparison, particularly with respect to competitors (Bathelt & Schuldt, 2008).

Thus, on different occasions, and through different pathways, global information about new and emerging industry or market trends circulates among the agents who are temporarily gathered at trade fairs. These agents benefit from recurrent, often short, face-to-face encounters with other trade fair participants, which in turn foster a specific communication and information ecology that has been defined in recent work as "global buzz" (Maskell et al., 2006). As explained more fully elsewhere (Bathelt & Schuldt, 2010),[7] global buzz is a complex and multidimensional concept that enables distinctive processes of knowledge creation and exchange through processes of interactive learning and learning-by-observation. Trade fairs bring together both leading and less well-known agents from an entire industry or technology field mainly to exchange knowledge about current and future developments in their industry. What is unique to the trade fair setting is that these exchanges take place around physical displays of products, prototypes, and innovations (Bathelt et al., 2014).

Trade fairs are usually organized according to a specific technological or industry focus. They are, thus, able to cut across distinct capitalist varieties and enable meaningful forms of interaction among actors from different national political economies. Indeed, the firms that participate at these events come from all over the world, operate in distinct institutional environments, use at least partially different technologies, focus on different markets, and apply diverse practices and processes. International trade fairs represent a particular setting wherein economic agents can acquire an overview of new developments and trends in the world market (Rosson & Seringhaus, 1995; Sharland & Balogh, 1996; Godar & O'Connor, 2001; Prüser, 2003). Exhibitors and visitors also benefit from the opportunity to inspect other exhibits at the trade fair and observe the visual representation of other firms' corporate cultures and strategies.

Although the composition of participants at international trade fairs tends to be fairly heterogeneous, recent research has found that different, partially related communities often find each other at these events (Entwistle & Rocamora, 2006). Interaction with other trade fair exhibitors helps to minimize the level of complexity and uncertainty associated with rapidly changing product and technology markets. Within their contact networks, economic actors are linked in different ways that allow them to exchange facts, impressions, gossip, and small talk. This provides participants with an opportunity to discuss their own experiences with existing and new products or processes in understandable ways. Relating different types of business knowledge and other information also gives trade fair participants a chance to investigate other actors and establish initial communication that can be followed up in the future (Borghini et al., 2006; Bathelt & Schuldt, 2008). Regular attendance at international trade fairs helps firms identify suitable business partners to complement their needs, establish trust with distant partners, and initiate the development

[7] The critical ingredients of "global buzz", as described by Bathelt and Schuldt (2010), are: (i) the dedicated co-presence of global supply and demand; (ii) intensive temporary F2F interaction; (iii) multiple possibilities for observation; (iv) intersecting interpretative communities; and (v) multiplex meetings and relationships. Central to these processes are integrational and informational cues (Short et al., 1976), which include verbal and non-verbal cues, visual stimuli, and feelings or emotions, all of which are characteristic of the communication and observation processes observed at international trade fairs.

of robust inter-firm networks in areas such as research, production, and marketing. Within this unique context of interaction and exchange, often across distinct geographic and political-economic lines, the types of discussions and knowledge flows taking place can serve to link markets and technologies with institutional and political conditions. I would argue that there are few, if any, other settings or occasions where such linkages are being created in quite as direct a way.

As the above discussion has demonstrated, the work of economic geographers has been particularly instrumental in developing a knowledge-based perspective of trade fairs (Bathelt et al., 2014). Conceptualized as temporary clusters, this work provides a strong theoretical foundation for the continued analysis of trade fairs in the context of global political economic patterns and processes.

2.2.6 Trade Fairs in Political Science

As previously mentioned, trade fairs have received scarce attention in the field of political science. In the political economy literature, for example, trade fairs, if mentioned at all, generally only receive passing reference. And, such references tend to be historical in nature, noting, for instance, the fairs of the medieval period or the world fairs (Frieden, 2006; O'Brien & Williams, 2016). Trade fairs in their contemporary form are barely ever mentioned in this literature. Indeed, many of the key textbooks in the field of political economy, including those with a global or international emphasis, do not mention trade fairs at all and do not list them in the index (see, for example, Gilpin, 2001; Ravenhill, 2008; Oatley, 2016). Frieden's work (2006) is an exception in that it devotes several pages to trade fairs; yet, as touched upon above, his focus is historical, describing the world fairs – e.g., London's Great Exhibition of 1851 and the Paris International Exhibition of 1900 – in the context of scientific and technological progress and the shifting landscape of industrial leadership at the time. In a more fleeting way, O'Brien and Williams' (2016) text also mentions trade fairs in an historical perspective, referring to a series of Flemish fairs in relation to the historical development of a commercial network in northern Europe. More specifically, it notes that the Belgian city of Bruges was a participant in these events by 1200, along with other indicators of its emergence as "a second Venice" (O'Brien & Williams, 2016, 47). Without denying the value of descriptive historical work, the above examples hardly exemplify a serious interrogation into the nature of trade fairs, let alone an exploration of their contemporary form and potential significance in shaping global economic processes today.

With respect to the relatively few studies in political science that refer to trade fairs in a less historical way, these events are usually mentioned in an incidental manner as part of a list, for instance, of different commercial activities diplomats engage in (Lee & Hudson, 2004); as evidence of trade gains within a national state – that is, when trade fair participation is high in that country (Feiler, 1998); or in the context of describing a country's broader economic development strategy or trade boosting strategy (Soludo et al., 2004). In general, this work speaks of trade fairs in a descriptive and transitory way. In other words, there is no acknowledgement of, or attempt to present these events as an integral part of the global political economy, let alone an investigation into the nature and extent of their impact on macroeconomic processes.

One notable exception to this trend in the political science literature is found in the work of international relations scholar John Gerard Ruggie. In a seminal article that explores how a particular form of territoriality – "disjoint, fixed, and mutually exclusive" – came to define modernity in international politics, Ruggie (1993, 155) has the following to say about trade fairs:

> In no sense could the medieval trade fairs have become substitutes for the institutions of feudal rule. Yet, the fairs contributed significantly to the demise of feudal authority relations. They did so because the new wealth they produced, the new instruments of economic transactions they generated, the new ethos of commerce they spread, the new regulatory arrangements they required, and the expansion of cognitive horizons they effected all helped to undermine the personalistic ties and the modes of reasoning on which feudal authority rested.

To be clear, trade fairs are not the main subject of Ruggie's (1993) work, nor does he refer to the contemporary model of trade fairs that are the specific focus of this book. Yet, what is particularly interesting in the above passage is Ruggie's observation that trade fairs can serve as a crucial source of economic dynamism. This acknowledgement – however small or isolated it may appear in the broader context of the field – has been largely overlooked. Indeed, it is not sufficient to stop at the observation that trade fairs "contributed significantly to the demise of feudal authority relations" (Ruggie, 1993, 155). Certainly, a much deeper investigation needs to be undertaken in order to understand the potential contribution of trade fairs with respect to shifts in institutional practices, for example, or the precise mechanisms by which these events might influence the economic development, growth, and competitiveness of national political economies.

It is in this context that I seek to encourage a greater focus on international trade fairs in the field of political science. Some examples of their potential significance to political science include the following: Trade fairs serve as a pivotal platform for the direct exchange of ideas, information, and knowledge among actors from different political and institutional regimes; these events can be used as a policy to stimulate economic development, with different trade fair policies used depending on the nature and development stage of the economy in question; trade fairs can also trigger the development of new technology fields; and trade fairs represent an important vehicle for discussing and evaluating the consistency of political solutions between countries – through, for example, presentations at these events. These are just a few areas that highlight the relevance of trade fairs and why it is important to extend the study of these events into the realm of political science, generally, and comparative political economy more specifically.

2.2.7 Summary of Key Findings from the Trade Fair Literature

The above review of the literature on international trade fairs suggests that these events have become increasingly important venues through which critical knowledge regarding global technologies, standards, and practices are exchanged between firms – including those that belong to different national economic systems. It therefore seems entirely plausible that leading international trade fairs might play a role in enabling processes of cross-national convergence with respect to new technologies, strategic choices, and industry "best-practices". There are, at least, three different ways that international trade fairs might accomplish this: First, by contributing to technological "ubiquitification" (Maskell & Malmberg, 1999a) – that is, the processes by which previously localized firm capabilities are transformed into ubiquities that are globally available at roughly the same price –[8] trade fairs have the potential to facilitate or accelerate such processes through both intended and unintended knowledge transfers and the signing of sales contracts; second, by supporting the diffusion of "best-practices", as firms from different parts of the world are brought together to scrutinize leading-edge firms and products, and by providing critical information about promising markets and successful business concepts; and, finally, by helping negotiate institutional differences at various levels, including exchanges between firms from different contexts and the dissemination of ideas about new national regulations and conditions for innovation.

In this study, I argue that international trade fairs represent crucial nodes in the global political economy that serve as important spaces within which cross-national differences in institutional and technological conditions or practices are negotiated. In this way, international trade fairs may be viewed as a key mechanism by which distinct national specializations in products and/or processes may converge. Yet, drawing key insights from the literatures on comparative capitalisms (see Chap. 3) and the garbage-can model of organizational choice (see Chap. 4), I suggest that inter-firm knowledge flows at trade fairs will not inevitably lead to processes of convergence with respect to the technological development patterns of cross-national firms. Although it is argued that trade fairs embody decisive mechanisms linking economic agents (i.e., firms) to broader national development patterns (i.e., technological standardization/specialization), the empirical chapters of the study

[8] Traditional location theory (Weber, 1909) makes a distinction between two different kinds of production inputs: (i) "localized materials", the costs of which differ dramatically across geographic location, and (ii) "ubiquitous materials" or "ubiquities", which are available across the globe at roughly the same price. Maskell and Malmberg (Maskell & Malmberg, 1999a; b) discuss the continued relevance of this Weberian distinction despite ongoing changes to what are considered the most important location factors (see, Maskell & Malmberg, 1999a). They also identify three processes driving change in the relative significance of location factors. Two of the processes are traditional: supply and demand; the third process is more recent: globalization. They provide several examples of how the process of globalization "*actively converts formerly localized inputs into ubiquities*" (Maskell & Malmberg, 1999a: 12 [emphasis original]), including the erosion of most trade barriers and the diminished loyalty of customers to local suppliers.

will demonstrate that the search behaviour of firms at these events is not typically guided by a linear or rational decision-making process. And, yet, despite the seemingly random and loosely structured nature of these searches, there does appear to be an overarching framework within which firms operate. I suggest that this framework closely relates to the capitalist system and its institutional infrastructure, thereby pointing more towards continued divergence in the specialization of firms from different capitalist varieties than convergence on a single model. These arguments will be developed much more fully in Chaps. 5 and 6.

2.3 Trade Fairs as a Source of Economic Dynamism

As indicated in the above literature review, trade fairs have been an important source of economic dynamism since at least the Middle Ages (Face, 1958; Ruggie, 1993). In their contemporary form, however, trade fairs may serve as increasingly significant drivers of economic growth in the context of the knowledge-based global economy. This is due to their role in temporarily bringing together agents from many different parts of the world and enabling knowledge flows regarding the global state of an entire industry and related market trends in an incredibly dense and intensified fashion (Maskell et al., 2006; Bathelt & Schuldt, 2008; Bathelt et al., 2014). With this in mind, I develop the argument that the modern international trade fair might represent a leading source of "knowledge-driven economic dynamism" – a term used by Arvanitidis and Petrakos (2011, 25) to denote "the potential an area has for generating and maintaining high rates of economic performance due to its knowledge capacity." Following Arvanitidis and Petrakos (2011), four key building blocks of knowledge-driven economic dynamism can be identified: (i) human capital, which refers to a well-educated and skilled workforce; (ii) innovation ability; (iii) information access; and (iv) economic performance.

Drawing on the work of Arvanitidis and Petrakos (2011) and other relevant literature, my specific interest in this part of the chapter is to address the outstanding question of what sets trade fairs apart from other key sources of economic dynamism. The answer, simply put, is that trade fairs encompass all four building blocks of the knowledge-driven economic dynamism construct. Indeed, by straddling or bringing together the various individual sources of economic dynamism, trade fairs may be viewed as a central mechanism in harnessing global knowledge flows across different economic systems. As this study argues, these knowledge flows may ultimately play a decisive role in the future development trajectories of different national political economies, specifically with respect to patterns of industry specialization or standardization. The remainder of this chapter begins to lay out the theoretical foundation for this argument through a stylized review of the literature on key sources of economic dynamism.

2.3.1 Traditional Determinants of Economic Dynamism

Over the last three decades or so, there has been a vast amount of literature focused on identifying the core determinants of economic growth or dynamism. Yet, the precise mechanisms underlying economic progress remain, at best, only partially understood and are often lacking in conceptual clarity. Although the growth literature is characterized by a diversity of theoretical perspectives and methodological approaches, at least two dominant schools of thought can be identified – (i) the neoclassical growth model inspired by Solow's (1956) growth model, which focused on the role of investment, in particular; and (ii) the more recent theory of endogenous growth following the work of Romer (1986) and Lucas (1988), which highlights factors such as human capital and innovation capacity. Other significant contributions to the growth literature include cumulative causation theory (Myrdal, 1957) and the new economic geography school (Krugman, 1991; Fujita et al., 1999). Each theory stresses the importance of different factors in determining economic progress. This literature may be further divided into studies that focus on "proximate" and "fundamental" determinants of economic development, where the former refers to the accumulation of capital, labour, and technology in economic progress, and the latter refers to institutions, political systems, socio-cultural factors, demography, and geography (see Rodrik, 2003; Snowdon, 2003; Acemoglu et al., 2005; Artelaris et al., 2011). Consequently, a broad range of factors have been advanced as potential sources of economic dynamism, with little sense of an overarching conceptual framework (van Hemert & Nijkamp, 2011).[9]

Without engaging in a lengthy discussion of the theoretical and methodological issues surrounding the literature on economic growth, the following section briefly reviews some of the main determinants of economic dynamism that have been identified in the literature.[10] This is done specifically with a view to positioning international trade fairs as a critical, albeit largely overlooked, source of economic dynamism in the context of the global knowledge economy. Throughout the course of this discussion, it should also become clear that international trade fairs are a phenomenon that deserves a much more intensive examination within the field of political economy research. The review presented below is structured according to proximate and fundamental sources of economic dynamism (Petrakos & Arvanitidis, 2008).

[9] Recently, however, an international project entitled "Dynamic Regions in a Knowledge-driven Global Economy" (DYNREG) has taken the first steps towards developing a unified general theory of economic dynamism. For more on this project, see: http://cordis.europa.eu/result/rcn/47101_en.html

[10] For a more detailed literature review on this topic, see Petrakos and Arvanitidis (2008).

2.3.2 Proximate Determinants of Economic Dynamism

Investment is widely recognized as one of the most critical factors affecting economic performance in both the neoclassical and endogenous growth model literatures, though the latter views investment as having more permanent effects than the former (Arvanitidis & Petrakos, 2011; Artelaris et al., 2011). Following its emphasis in the theoretical literature, there has been a large amount of empirical research examining the relationship between investment and economic growth. Yet, few studies have been able to demonstrate a clear causal linkage between investment and economic progress (see, for example, Kormendi & Meguire, 1985; De Long & Summers, 1991; Mankiw et al., 1992; Auerbach et al., 1994; Barro & Sala-i-Martin, 1995; Podrecca & Carmeci, 2001).

Human capital is regarded as a primary source of economic dynamism in a number of endogenous growth models (see, for example, Lucas, 1988). We will revisit this determinant again in the section on knowledge-based economic dynamism, as it represents one of the four essential elements of this concept (Arvanitidis & Petrakos, 2011). However, it is interesting to note that while the term human capital embodies a variety of social and health factors, much of the literature concentrates specifically on the education variable, including, for instance, school enrolment rates, adult literacy rates, highest level of educational attainment, and, more recently, international mathematics and science test scores (Petrakos & Arvanitidis, 2008). Whereas many scholars have discovered a significant positive relationship between education and growth (see, for instance, Barro, 1991; Mankiw et al., 1992; Barro & Sala-i-Martin, 1995; Hanushek & Kimko, 2000), others have questioned the explanatory strength of this variable (see, for example, Benhabib & Spiegel, 1994; Pritchett, 2001).

Innovation and/or R&D activities are strongly linked to economic productivity and growth due, in part, to contemporary society's increasing reliance on technology, which, in turn, supports the introduction of new product or process innovations (Petrakos & Arvanitidis, 2008). Innovation activity, like knowledge flows, is often measured using patents or patent applications as a proxy, though investments in R&D have also been used quite frequently. These traditional measures are limited and provide, at best, only a partial understanding of the nature and extent of innovation activities. That being said, a large body of empirical work supports the relationship between innovation and R&D activities and economic progress (see, for instance, Fagerberg, 1987; Lichtenberg, 1992; Ulku, 2004). On this point, it is not difficult to see how international trade fairs could support the potential impact of innovation activities on economic growth. The presentation and observation of new technologies and innovations are widely considered to be pivotal aspects of these events.

While the literature on proximate factors has revealed many important insights regarding potential sources of economic growth and cross-national income disparities, it has been criticized for failing to adequately address the deeper determinants of economic growth and development. According to North and Thomas (1973, 2 [emphasis original]), "[t]he factors we have listed (innovation, economies of scale, education, capital accumulation, etc.) are not causes of growth; *they are growth*". If, for instance, technology, human capital, and physical capital are at the core of differences in income across countries, one must ask why some countries focus more on these aspects than others? In other words, a singular focus on proximate factors offers an incomplete narrative. To obtain a more complete understanding of the problem, scholars have thus become increasingly focused on more fundamental determinants of economic dynamism – a topic to which I now turn.

2.3.3 Fundamental Determinants of Economic Dynamism

Institutions represent an important "deep" source of economic dynamism. Although the relationship between institutions and economic growth is not a new discovery (see, for instance, Ayres, 1962), in recent years we have seen renewed interest in such factors and more systematic empirical attempts to investigate the precise impact of institutions on economic performance (see, for example, Knack & Keefer, 1995; Rodrik, 1999; Acemoglu et al., 2002). Institutions have been variously defined, and such definitions often discussed at length, in different literatures (see, for example, March & Olsen, 1989; Powell & DiMaggio, 1991; Granovetter & Swedberg, 1992; Steinmo et al., 1992; Koelble, 1995; Helmke & Levitsky, 2004; Bathelt & Glückler, 2014). For present purposes, however, institutions can be defined as both the formal (e.g., constitutions, laws) and informal (e.g., customs, codes of conduct, sanctions) constraints that structure political, economic, and social interaction (North, 1990).

A number of institutions have been identified as playing a key role in economic growth. Rodrik (1999; see also, Rodrik et al., 2002), for instance, identifies five different types of institutions that support economic development: (i) property rights, (ii) regulatory institutions, (iii) institutions for macroeconomic stabilization, (iv) institutions for social insurance, and (v) institutions for conflict management. Moreover, over the last couple of decades, a growing body of empirical work, employing a variety of different variables, has demonstrated that there is a strong relationship between the quality of the institutional framework and the nature of economic growth (see, for example, Knack & Keefer, 1995; Mauro, 1995; Rodrik, 1999; Rodrik et al., 2002; Acemoglu & Johnson, 2005). Following North's (1990) definition, trade fairs may be viewed as an institution, as they provide a framework within which productive action and interaction between economic agents may be carried out. This study suggests that international trade fairs represent an additional, but still largely under-explored, type of market-supporting institution.

Political factors have also received considerable attention in the academic literature for their role in shaping economic growth. The substantial amount of research in this area suggests that political issues, such as regime stability and level of democracy, greatly affect economic progress (Lipset, 1959; Kormendi & Meguire, 1985; Scully, 1988; Grier & Tullock, 1989). Different aspects of the political environment have been explored in an effort to determine the nature of this relationship, including government stability, political violence, and subjective perceptions of politics (Brunetti, 1997). While such political factors might seem far removed from international trade fairs, the growth of the trade fair industry in regions such as the Middle East may indicate a potential role for these events in decreasing the impact of negative political factors on economic growth.

The importance of *geography* in explaining questions related to economic growth, including cross-country variation in income, has long been acknowledged in the literature (see, for example, Gallup et al. 1999). A considerable body of empirical work has demonstrated that climate, endowment of natural resources, topography, transport costs, and the flow of knowledge and technology from more advanced regions directly influence economic growth, specifically in relation to agricultural productivity and human resources (Sachs & Warner, 1997; Bloom et al., 1998; Masters & McMillan, 2001; Armstrong & Read, 2004). In this context, the topic of trade fairs seems relevant, as these events have the capacity to reduce transportation costs and problems associated with knowledge and technology transfer at a distance. There has, however, been some skepticism regarding the explanatory power of geography in this context. For example, some scholars have found that once institutions are controlled for, geographical factors have minimal impact on specific issues of economic dynamism, such as income disparities across national economies (Rodrik et al., 2002; Easterly & Levine, 2003).

While it is useful to identify some of the main determinants of economic dynamism, it is important to recognize that these forces do not operate in a vacuum. Indeed, they often interact with one another in a relational way, making their individual influence on economic progress extremely difficult to discern. Recent efforts to develop a composite indicator of economic dynamism in the context of the knowledge-based economy represent a good first step in overcoming this challenge (Arvanitidis & Petrakos, 2011).

2.3.4 *Defining the Knowledge-Based Economy*

Whereas knowledge has always played a significant role in the economy, since the mid-1990s, in particular, there has been an observable increase in the extent to which knowledge can shape processes of economic and social development. Indeed, the term "knowledge-based economy" is now widely employed by scholars, government practitioners, the media, and the public at large (Dosi, 1995; David & Foray, 2002; Brinkley, 2006). This shift in the scale and importance of knowledge in economic processes has brought about new rules, practices, and institutions, which underpin and support the knowledge-based economy. Knowledge-based economies have been defined as "economies which are directly based on the production, distribution, and use of knowledge and information" (OECD, 1996, 3). In a knowledge-based economy, knowledge is, thus, considered to be both quantitatively and qualitatively more significant than it was in the past. Moreover, such an economy is largely driven by the use of information and communication technologies (ICTs) (Godin, 2006).

Scholars have identified at least three core shifts associated with the emergence of the knowledge-based economy (Soete, 2006; Arvanitidis & Petrakos, 2011). The first shift relates to the suggested commodification of knowledge (Drucker, 1998; OECD, 1999). From this perspective, knowledge is no longer viewed as a "black box"; instead, it is treated as a factor that is internal to the economic system and, thus, subject to economic principles associated with its production and exchange (Arvanitidis & Petrakos, 2011). But knowledge differs from traditional material commodities in at least two important ways – namely, price determination and knowledge transfer (Bathelt & Glückler, 2011). With respect to price determination, knowledge is unlike traditional commodities in that it is much more difficult to exchange through market transactions given the problems associated with determining its precise supply and demand. This seems to be true even with respect to the more explicit, codified forms of knowledge, as it remains difficult for a potential buyer to determine the exact value of such knowledge without having a clear understanding of its contents (Bathelt & Glückler, 2011). Moreover, it is argued that effective interpretation and integration of new explicit knowledge depends upon a sufficient base of tacit knowledge (Maskell & Malmberg, 1999b).

The reality of knowledge as a commodity is also impeded by the particular nature of knowledge exchange processes. Unlike traditional commodities, tacit knowledge, in particular, tends to be highly localized and embodied in people or technical systems (Polanyi, 1967; Maskell & Malmberg, 1999b; Amin & Cohendet, 2004; Bathelt & Glückler, 2011). All of this makes it extremely difficult to exchange in ways similar to material resources or products. As will be discussed in greater detail later on, trade fairs present an effective response to at least some of the challenges associated with the commodification of knowledge. One reason for this is because part of the knowledge base that is associated with the economic and institutional structure of a country becomes embodied in the firms that operate within those settings. These structures are integrated into the learning processes of firms and ultimately find representation in their products, services, and innovations. International trade fairs may be seen as a crucial platform for visualizing, discussing, and internalizing these representations.

The second identified shift involves the centrality of ICTs in the creation and dissemination of knowledge (Lundvall & Foray, 1998; Arvanitidis & Petrakos, 2011). From this perspective, ICTs facilitate the rapid collection, organization, storage, and dissemination of data, thereby accelerating and supporting processes of knowledge transfer (Roberts, 2000; Arvanitidis & Petrakos, 2011). Indeed, there is much evidence to suggest that ICTs have greatly increased communication flows among agents located at various distances from one another. They have aided the transferability of knowledge through, for instance, increased communication among scholars, enhanced opportunities for collaboration, and improved access to information, education, and training. Yet, ICTs favour highly codified and standardized forms of knowledge. As such, they are of questionable utility in relation to the more tacit forms of knowledge (Roberts, 2000). Other mechanisms must, therefore, be used to support the exchange of tacit knowledge – including, I will argue, international trade fairs.

The final shift in appreciating the changing role of knowledge in the economy involves the nature of innovation processes. The core of the argument here is that innovation capacity has come to depend on the extent to which firms are able to simultaneously combine and reorganize existing knowledge, as opposed to discovering entirely new technological principles (David & Foray, 2002; Arvanitidis & Petrakos, 2011). According to this interpretation, innovation processes may be less contingent upon research than they were before, but rely very heavily on access to the latest technologies and established procedures for disseminating such knowledge (Arvanitidis & Petrakos, 2011). As will be demonstrated below, leading international trade fairs are uniquely positioned to respond to this observed shift in that they provide unparalleled access to the latest technological advancements in a specific industry or field. They also create an environment that provides a fairly clear set of expectations and practices regarding the exchange of such knowledge during the actual trade fair – although considerable knowledge exchange can occur through unexpected or unplanned processes that take place outside the actual fairgrounds (e.g., restaurants, bars, cafes).

2.3.5 Knowledge-Based Economic Dynamism

As mentioned above, the term knowledge-based economic dynamism involves four key building blocks: (i) human capital, (ii) innovation capacity, (iii) information access, and (iv) economic performance. Informed especially by the work of Arvanitidis and Petrakos (2011) and Artelaris et al. (2011), a brief description of each follows below.

Simply put, *human capital* refers to a skilled and well-educated workforce. This is widely understood as a key ingredient in the successful acquisition and utilization of existing knowledge, but it is also instrumental in the creation and dissemination of new knowledge. While human capital includes a variety of dimensions, as mentioned above, empirical studies have largely focused on education as the main indicator of human capital. Higher education, in particular, has been strongly associated with the production of new knowledge and the creative use or recombination of existing knowledge (Artelaris et al., 2011; Barro & Sala-i-Martin, 1995). In previous work on the determinants of economic growth, human capital is considered a major source of economic dynamism, particularly from the perspective of endogenous growth models (Romer, 1986; Lucas, 1988; Artelaris et al., 2011).

Attending or participating in trade fairs may be viewed as another form of human capital. As Donckels and Lambrecht (1995) point out, participation in these events leads to the accumulation of human capital by allowing entrepreneurial firms to explore new markets and resources. Trade fairs are also a venue in which highly successful or entrepreneurial firms can provide assistance to newer, less established firms. Further, trade fairs are important sites for technological learning and may, thus, be considered as a support structure for the educational variant of human capital.

Innovation capacity is the second core component of the knowledge-based economic dynamism indicator. It refers to the establishment of a strong and stable innovation system, which can occur at different levels (e.g., national, regional, technological, or industrial) and is comprised of firms, research centres, and other related or supporting organizations and institutions (Arvanitidis & Petrakos, 2011). According to several studies in the endogenous growth literature, the increasing importance of innovation and R&D activities to economic development is directly related to the larger role technology now plays in bringing about new products and processes. As noted above, international trade fairs play a key role in enhancing innovation capacity. Indeed, the stimulation and promotion of innovation is often recognized as one of the central functions of these events.

Information access is the third critical component of knowledge-based economic dynamism. It relates to the use of ICTs. As previously mentioned, ICTs have had a profound impact on patterns of communication and the spread of information across all parts of the world. Several studies have been conducted on the linkages between ICTs and economic growth. This research seems to indicate that ICTs have become a key ingredient of knowledge-based economic dynamism, as both ICT usage and the development of ICT sectors have a positive impact on economic productivity (see Colechia & Schreyer, 2002; Avgerou, 2003). Again, there is an obvious relationship between this building block and trade fairs. As articulated in the knowledge-based view of these events, international trade fairs are home to a unique information and communication ecology known as "global buzz" (Bathelt & Schuldt, 2010). They are events that create many opportunities for firms to systematically obtain information about the technical and strategic choices of a variety of economic actors, including competitors, customers, and suppliers: "On different occasions, and through different routes, global information about trends and ideas in an industry, as well as all sorts of news and gossip, flow back and forth between the participants who are temporarily clustered at trade fairs" (Bathelt & Turi, 2013, 68).

Economic performance represents the final aspect of knowledge-based economic dynamism. The main argument here is that existing economic conditions can profoundly influence an economy's ability to create and make effective use of knowledge as a key tool for economic growth (Arvanitidis & Petrakos, 2011). More specifically, weak economic foundations can impede knowledge-driven economic dynamism, while a robust economic base can support it (Arvanitidis & Petrakos, 2011). The hypothesis that past economic performance shapes future economic growth is well explored in the literature, especially with respect to questions of cross-national economic convergence or divergence (see Baumol, 1986; Grier & Tullock, 1989; Barro, 1991; Barro & Sala-i-Martin, 1995; Sala-i-Martin, 1996). This body of work provides strong evidence to suggest that past economic conditions play a pivotal role in shaping future economic dynamism. As such, it is a testament to the power of path-dependent processes in patterns of economic development.

Trade fairs correspond strongly to the four components of knowledge-based economic dynamism described above. With respect to innovation capacity and information access, the relationship is quite obvious. However, trade fairs also relate quite closely to the other two elements: human capital and economic performance. With respect to human capital, trade fairs may be viewed as critical sites of technological learning. This educational function can be seen in the form of organized innovation competitions, expert-led seminars, and other scheduled educational opportunities; it can also be seen less formally through in-depth discussions and demonstrations with engineers and technical experts at exhibitors' booths. With respect to economic performance, trade fairs take on more of a mediating role. On the one hand, the circulation of information, ideas, and knowledge at these events could be potentially disruptive to the path dependent processes discussed above. On the other hand, the nature of the interactions among agents at these events may serve to reinforce existing trajectories. These two potential outcomes are of particular relevance to the empirical puzzle underlying this study as well as the convergence/divergence debate, the latter of which will be addressed more fully in the next chapter.

Within the context of the broader literature on determinants of economic growth, the concept of knowledge-based economic dynamism may be accurately defined as a proximate rather than a fundamental cause of economic growth. Yet, as previously indicated, leading international trade fairs encompass more than the four building blocks identified by Arvanitidis and Petrakos (2011). They also include, for example, geography, institutions, and human capital. Thus, I argue that while the concept of knowledge-based economic dynamism is useful in demonstrating the complex and multi-faceted nature of economic dynamism in today's economy, major international trade fairs are, at once, both proximate and fundamental in their orientation. It is their capacity to straddle both forms of economic growth factors and connect multiple determinants simultaneously that makes them a unique and potentially potent source of economic dynamism in their own right.

2.4 The Trade Fair Sites Used in the Study

Having discussed how trade fairs have been addressed in different literatures as well as their potential significance as a source of economic dynamism in the global knowledge economy, this section now turns to the specific trade fairs sites used in this study. It is important to note that my focus in this research is exclusively on business-to-business (B2B) events, rather than business-to-consumer (B2C) events. This is due to my particular interest in the technological search and knowledge processes of producers, as opposed to consumer-based choices or consumer behaviour.

For the quantitative analysis, the empirical setting consisted of four international trade fairs: *bauma* (Germany), *Electronica* (Germany), *EMO Hannover* (Germany), and the *International Consumer Electronics Show* (CES) (United States). According to the knowledge-based typology of trade fairs developed, first, by Golfetto (2004) and, later, by Bathelt et al. (2014), these four fairs fit, or come very close to fitting, the category of "Hub Trade Shows", where the majority of participants come from overseas and the trade fair's primary focus is on international exchanges. Beyond this, these four trade fairs include two broad types of industry/product groups: (i) *investment goods*, with a particular emphasis on machinery products; and (ii) *consumer goods*, which have more of a technical or design focus.[11] My decision to include different sectors/industries in the study was based on the view that different institutional arrangements promote different types of innovative competencies that may influence spatial patterns of technological specialization (Lam & Lundvall, 2006). Moreover, as Lam and Lundvall (2006) observe, learning and technological opportunities occur in all parts of the economy. In fact, "learning taking place in traditional and low-tech sectors may be more important for economic development than learning in a small number of insulated high-tech firms" (Lam & Lundvall, 2006, 135). Finally, all four trade fairs are recognized as leading international events in respected trade fair industry/association reports, journals, and websites (e.g., AUMA, TSEA, Imp-Exp Executive Magazine).

The dominant characteristics of the four trade fairs studied in the quantitative analysis are summarized below in Table 2.1. In the period under review, these four events had a combined total of 10,348 exhibitors[12] and 778,685 registered visitors. The strong international significance of these trade fairs is evident in the number of exhibitors coming from outside the host country.

As discussed in the previous chapter, the first two trade fairs – *bauma* and *Electronica* – were used exclusively to collect data through the online survey method, whereas the latter two trade fairs – *EMO* and the *International CES* – were used to collect data through both the online and in-person survey modes. A description of each of these four events follows at the end of this chapter.

For the qualitative part of the study, the empirical setting consisted of two trade fairs in the United States – *LightFair International (LFI)* and *International Plastics Showcase (NPE)* – as well as two trade fairs in Canada – *Canadian Manufacturing Technology Show (CMTS)* and *Montreal Manufacturing Technology Show (MMTS)*. As with the trade fairs studied in the quantitative analysis, these four North American trade fairs include the same two broad types of industry/product groups: *investment goods* and *consumer goods*. Although the four North American fairs are not equal to each other in terms of their international significance (this has a bearing on the diversity of firms attending the trade fair and, consequently, the types of partners that may be possible), all of the events are regarded as leading B2B events in their respective industries in North America. Practical considerations, including the timing of the events and their geographic location also played a role in the selection of these trade fairs. The main characteristics of the four North American trade fairs studied in the qualitative analysis are displayed in Table 2.2. In the period under

[11] This categorical distinction follows the work of Schuldt and Bathelt (2011), which uses the investment goods/consumer goods dichotomy to classify seven international trade fairs in the German context.

[12] As will be noted later on, this number is larger than the total number of exhibitors included in the survey distribution list. This is because some of the firms listed on the trade fair websites did not include electronic addresses.

Table 2.1 Characteristics of the four international trade fairs studied in the quantitative analysis, Germany and the USA

Trade fair	Year	Rhythm of events	Location	Product focus	Exhibitors (number)	Visitors (number)	Foreign Exhibitors (number of countries)	Exhibition space (m^2)
Bauma	2010	Triennial	Germany	Investment goods	3150	420,000	53	555,000
Electronica	2010	Biennial	Germany	Consumer goods	2578	72,185	45	142,500
EMO	2007[a]	Biennial	Germany	Investment goods	2120	166,500	42	180,158
International CES	2011	Annual	USA	Consumer goods	2500	120,000	38	158,000

Sources: Messe Mūnchen GmbH (2010a, b), Imp-Exp Executive Magazine (Sail by Exhibition, 2010). Top 100 World Trade Fairs.

[a]The data were not available for EMO 2009 during the data collection period for this study. Consequently, I relied on data from EMO, 2007 to develop the exhibitors list, which was then used for the distribution of the online survey

review, these four events had a combined total of 3028 exhibitors and 78,314 registered visitors. Although the international importance of these trade fairs overall is considerably less than the events selected for the quantitative analysis, as mentioned above, all four of the North American trade fairs represent leading events in their respective industries in the country in which they are hosted.

The four North American trade fairs selected for the qualitative analysis have a number of notable advantages. First, by choosing fairs that can be categorized according to their broad orientation or focus – in this case, design-oriented fairs and manufacturing-oriented fairs – we can make comparisons based on the investment goods/consumer goods dichotomy noted above (Bathelt & Schuldt, 2008). It is possible to explore, for instance, whether the search behaviours of exhibiting firms at design fairs differ from that of exhibitors at manufacturing fairs in meaningful or systematic ways. Second, the selected trade fairs demonstrate markedly different settings for innovation. For instance, in the past decade or so, the lighting industry underwent a new surge of innovations, resulting in opportunities for a wide range of design innovations following changing environmental regulations – e.g., the phasing out of the classical light bulb (Bathelt et al., 2014; Bathelt & Gibson, 2015). The manufacturing technology fairs (i.e., NPE, CMTS, and MMTS) revealed a more stable setting for innovation than the trade fair focused on the lighting industry (i.e., LFI). In this latter context, the industry was more diverse in its focus and innovation appeared slower and more incremental in nature. This is relevant given the observed differences in innovation patterns across LMEs and CMEs (Hall & Soskice, 2001), which will be discussed later on in the book. Third, by selecting events that correspond to different trade fair types (Golfetto, 2004; Bathelt et al., 2014), it is possible to investigate variations in the nature of firm interactions at smaller, regional/national events (CMTS and MMTS) versus larger, more international fairs (LFI and NPE). And, finally, selecting cases from both the United States and Canada allowed me to make cross-national comparisons.

Before describing each of the trade fairs in greater detail, it is important to underline the relationship between search and innovation, as the focus of firms' search

Table 2.2 Characteristics of the four North American trade fairs studied in the qualitative analysis

Trade fair	Year	Rhythm of events	Location	Product focus	Exhibitors (number)	Visitors (number)	Foreign Exhibitors (%)	Exhibition space (m^2)
LFI	2009	Annual	USA	Consumer goods	476	23,416	N/A	170,000
NPE	2009	Triennial	USA	Investment goods	1851	44,000	33	950,000
CMTS	2009	Annual	Canada	Investment goods	500	5898	4	N/A
MMTS	2010	Biennial	Canada	Investment goods	201	5000	1	N/A

Notes: *N/A* not available, *LFI* LightFair International, *NPE* International Plastics Showcase, *CMTS* Canadian Manufacturing Technology Show, *MMTS* Montreal Manufacturing Technology Show behaviour at trade fairs has invariably been linked in some way to the discovery of new innovations. Yet, what constitutes "innovation" can differ substantially from one firm to the next. For example, in some contexts, customized production itself may be viewed as a kind of innovation. Moreover, innovations take shape through remarkably different pathways and such pathways are often associated with different institutional support configurations – e.g., more rapid innovations are often linked to LMEs, whereas more incremental innovations are often seen as a product of CMEs (Hall & Soskice, 2001; for a counter to this, see Taylor, 2004). While the different trade fairs used in the qualitative analysis tend to feature presentations of either more stable and incremental technologies (e.g., NPE) or more rapid and wholesale technological shifts (e.g., LightFair), I am particularly interested in the political-economic contexts of the firms exhibiting at these two types of events, rather than comparing the individual events themselves.

2.4.1 Trade Fairs Studied in the Quantitative Analysis

The trade fair entitled *bauma* is recognized as the world's largest and most significant event for the building and construction industry. The original bauma was established over 50 years ago and is now held every 3 years in Munich, Germany.[13] Despite a ban on air travel in Europe, which prevented many overseas visitors and approximately 50 foreign exhibitors from participating in the fair, bauma, 2010 attracted nearly 420,000 visitors from more than 200 different countries and 3150 exhibitors from 53 different countries (bauma, 2010). Held at the Neue Messe München exhibition centre (New Munich Trade Fair Centre), bauma, 2010 occupied 555,000 m^2 of space. In terms of its international significance, 60% of exhibitors and 35% of visitors came from countries outside Germany, in 2010 (bauma, 2010). bauma also has a strong reputation for the presentation of new innovations.

Electronica, in Munich, Germany, focuses on the complete range of electronics products and services. It is regarded as the world's leading trade fair for electronic components, systems, applications and services for all sectors. A biennial event, Electronica is considered a hub trade fair (Golfetto, 2004; Bathelt et al., 2014) with its clear focus on the international marketplace. In 2010, the year that I gathered the exhibitors list for the online survey, Electronica attracted 2578 exhibitors from 45 different countries and 72,185 visitors from 76 different countries. To put this into greater perspective, 59% of exhibiting firms and nearly 50% of registered visitors came from countries outside of Germany. Like bauma, Electronica takes place at the Neue Messe München exhibition centre (New Munich Trade Fair Centre). In 2010, it occupied approximately 142,500 m^2 of exhibition space (Messe München GmbH, 2010a, b).

[13] Different branches of bauma have since been developed, including bauma China, which takes place every 2 years in Shanghai, and bauma ConExpo in India, which was held for the first time in Mumbai in 2011.

EMO Hannover (EMO) is the world's largest trade fair for metalworking technology and is considered an important meeting place for the entire international sector. This trade fair presents a wide range of products and services throughout all production areas, including cutting and forming machine tools, manufacturing systems, precision tools, automated materials, computer technology, as well as industrial electronics and accessories. EMO follows a biennial trade fair rhythm, where it is held twice consecutively in Hannover, followed by once in Milan (e.g., Hannover – Hannover – Milan – Hannover – Hannover – Milan). With exhibitors coming from all over the world, EMO Hannover is a major platform for the entire metalworking industry. In 2007, (the year from which the exhibitors list was drawn for the online survey), EMO had 2120 registered exhibitors from 42 different countries and 166, 500 visitors. Taking place on the Hannover Fairgrounds, EMO, 2007 occupied a total of 180,158 square meters of exhibition space. With 60% of exhibitors and 36% of visitors coming from outside of Germany, EMO has a strong international character and a clear focus on engaging the international market (EMO, 2007, 2011).

The *International Consumer Electronics Show (CES)* is recognized as the world's largest international trade fair for consumer technologies and the largest trade fair of any kind in North America. Following an annual trade fair rhythm, CES runs for a period of 4 days. It is held at the Las Vegas Convention Centre in Las Vegas, Nevada, U.S. and, in 2011, occupied an exhibition space of 158,000 square meters. In 2011, (the year from which the exhibitors list was drawn for the online survey), CES attracted 120,000 registered visitors and 2500 exhibiting firms, more than 25% of which came from outside the U.S. Thus, while CES is still largely focused on international markets, its composition of overseas participants suggests less of an international or "hub" character compared to the other three events (CES, 2011).

2.4.2 Trade Fairs Studied in the Qualitative Analysis

The *International Plastics Showcase (NPE)* is recognized as a leading event for the plastics industry and one of the top five manufacturing trade fairs in the United States. Focusing on the latest in plastics and manufacturing technologies, NPE brings together the industry's entire supply chain from the ideation stage to materials and equipment to processers to brand owners. NPE is a technology-oriented fair that attracts a wide range of market segments, including the automotive, electronic, medical, chemical, and transportation industries. A triennial fair, NPE marked its 72nd anniversary in 2009 (the year the data were collected). Traditionally held at Chicago's McCormick Place, the 2009 event had 1851 exhibitors occupying over 950,000 square feet of exhibition space and an estimated 44,000 visitors. The NPE's international significance is quite strong, with registered visitors representing over 100 different countries and about one-third of exhibiting firms coming from outside of the U.S. (NPE, 2009).

LightFair International (LFI) is described as the largest and one of the most important events focusing on the architectural and commercial lighting industries in

North America. An annual trade fair, the location of LFI alternates every year between Las Vegas, Nevada and New York, NY. LFI is a mid-sized design-oriented fair. In the year the data were collected, 2009, LFI attracted close to 500 exhibitors and approximately 23,000 visitors (LightFair International, 2009a, b). In 2009, the trade fair was held at the Jacob K. Javits Center in downtown New York, in an exhibition hall of approximately 170,000 square feet. In terms of its international standing, registered visitors came from 73 different countries and all of the U.S. states, while roughly 20% of the exhibiting firms represented countries outside of the United States (LightFair International, 2009a, b).

The Canadian Manufacturing Technology Show (CMTS) is described as the largest manufacturing trade fair of its kind in Canada. A technology-oriented trade fair, CMTS adopts a broad focus on manufacturing technologies, including welding, robotics, optical solutions, as well as machine tools and applications. Taking place on an annual basis, the CMTS was held at Toronto's Direct Energy Centre, in 2009, the year the data were collected.[14] In that year, CMTS attracted 500 exhibitors and nearly 6000 registered visitors. CMTS lacks the international significance of the two U.S. fairs, with most (96%) visitors originating from within Canada and 90% from the province of Ontario. It is, thus, better classified as a regional/national trade fair, rather than an international or hub fair (CMTS, 2009).

Like its Ontario counterpart, the *Montreal Manufacturing Technology Show (MMTS)* has a broad manufacturing focus, with a particular emphasis on machine tools, metalworking, automation, and additive manufacturing. A biennial fair, MMTS runs for a period of 3 days at Place Bonaventure in downtown Montreal, Quebec. While MMTS is considered the leading trade fair of its kind for the North American French marketplace, it is comparatively small in size, with a very clear regional focus. In 2010, the year the data were gathered, it had approximately 200 exhibiting firms and 5000 registered visitors, most of which were drawn from within the province of Quebec. It is, thus, best classified as a regional/national event (MMTS, 2010).

2.5 Conclusion

Through a review of the existing literature on trade fairs, this chapter has sought to provide answers to the following questions: Why should we care about international trade fairs? How do they relate to the global political economy and to ongoing debates about institutional change? And what makes them different from other identified sources of economic dynamism? The literature review suggests that leading international fairs can serve as important temporary platforms for establishing trust with geographically distant business partners and developing durable inter-firm

[14] In more recent years, the CMTS has been held at the International Centre in Mississauga, Ontario.

networks in areas such as production, research, and marketing (Borghini et al., 2004; Maskell et al., 2006; Bathelt & Schuldt, 2008). Moreover, since trade fairs are typically organized according to a specific industry or technological focus, they cut across geographical and political boundaries and create numerous opportunities for F2F interaction and knowledge exchange among firms from different national economic systems. Such interactions may encourage firms to adopt new or different technological practices, eventually leading towards a shift in the national institutional bases of these firms. Finally, in the context of other identified sources of economic dynamism, including the fairly recently developed composite knowledge-based economic dynamism indicator, trade fairs emerge as a strong and uniquely multi-dimensional source of economic dynamism in their own right.

Against this backdrop, the following questions arise: how exactly do international trade fairs influence patterns of economic development and growth? What is the precise nature of the search and knowledge processes of firms during these events? And how might these processes eventually shape the development paths of firms with respect to future patterns of technological specialization or standardization at the national level? These questions will be addressed in the subsequent chapters. But before doing so, Chap. 3 continues to establish the theoretical foundations of the study by turning our attention to the literature on comparative capitalist variation.

References

Acemoglu, D., & Johnson, S. (2005). Unbundling institutions. *Journal of Political Economy, 113*(51), 949–995.

Acemoglu, D., Johnson, S., & Robinson, J. A. (2002). Reversal of fortune: Geography and institutions in the making of the modern world income distribution. *Quarterly Journal of Economics, 117*(4), 1231–1294.

Acemoglu, D., Johnson, S., & Robinson, J., (2005). Institutions as a fundamental cause of long-run growth. In P. Aghion & S. Durlauf (Eds.), *Handbook of economic growth* (Vol 1A). Elsevier.

Allix, A. (1922). The geography of fairs: Illustrated by old-world examples. *Geographical Review, 12*, 532–569.

Amin, A., & Cohendet, P. (2004). *Architectures of knowledge: Firms, capabilities, and communities*. Oxford University Press.

Armstrong, H., & Read, R. (2004). *The economic performance of small states and islands: The importance of geography*. Paper presented at Islands of the world VIII international conference: Changing islands – changing worlds.

Artelaris, P., Arvanitidis, P. A., & Petrakos, G. (2011). Explaining knowledge-based economic growth in the world economy. In P. Nijkamp & I. Siedschlag (Eds.), *Innovation, growth and competitiveness: Dynamic regions in the knowledge-based world economy* (pp. 41–59). Springer.

Arvanitidis, P. A., & Petrakos, G. (2011). Defining knowledge-driven economic dynamism in the world economy: A methodological perspective. In P. Nijkamp & I. Siedschlag (Eds.), *Innovation, growth and competitiveness: Dynamic regions in the knowledge-based world economy* (pp. 15–39). Springer.

Auerbach, A. J., Gokhale, J., & Kotlikoff, L. J. (1994). Generational accounting: A meaningful way to evaluate fiscal policy. *Journal of Economic Perspectives, 8*(1), 73–94.

Avgerou, C. (2003). The link between ICT and economic growth in the discourse of development. In M. Korpela, R. Montealegre, & A. Poulymenakou (Eds.), *Organizational information systems in the context of globalization. IFIP – The International Federation for Information Processes* (Vol. 126, pp. 373–386). Springer.

Ayres, C. E. (1962). *Theory of economic progress: A study of the fundamentals of economic development and cultural change*. Schocken.

Barro, R. J. (1991). Economic growth in a cross section of countries. *Quarterly Journal of Economics, 106*(2), 407–443.

Barro, R. J., & Sala-i-Martin, X. (1995). *Economic growth*. McGraw-Hill.

Bathelt, H., & Gibson, R. (2015). Learning in "organized anarchies": The nature of technological search processes at trade fairs. *Regional Studies, 49*(6), 985–1002.

Bathelt, H., & Glückler, J. (2011). *The relational economy: Geographies of knowing and learning*. Oxford University Press.

Bathelt, H., & Glückler, J., (2014). Institutional change in economic geography. *Progress in Human Geography, 38*(3), 340–363.

Bathelt, H., & Schuldt, N. (2008). Between luminaires and meat grinders: International trade fairs as temporary clusters. *Regional Studies, 42*(6), 853–868.

Bathelt, H., & Schuldt, N. (2010). International trade fairs and global buzz, part I: Ecology of global buzz. *European Planning Studies, 18*(12), 1957–1974.

Bathelt, H., & Turi, P. (2013). Knowledge creation and the geographies of local, global, and virtual buzz. *Knowledge and the Economy, 5*, 61–78.

Bathelt, H., Malmberg, H., & Maskell, P. (2004). Clusters and knowledge: Local buzz, global pipelines and the process of knowledge creation. *Progress in Human Geography, 28*, 31–56.

Bathelt, H., Golfetto, F., & Rinallo, D. (2014). *Trade shows in the globalizing knowledge economy*. Oxford University Press.

bauma. (2010, April 19–25). bauma 2010 Fact Sheet, Messe Mūnchen International.

Baumol, W. J. (1986). Productivity growth, convergence, and welfare: What the long-run data show. *The American Economic Review, 76*(5), 1072–1085.

Bautier, R.-H. (1970). The fairs of champagne. In R. E. Cameron (Ed.), *Essays in French economic history* (pp. 42–63). Richard D. Irwin.

Belizzi, J. A., & Lipps, D. J. (1984). Managerial guidelines for trade show effectiveness. *Industrial Marketing Management, 13*, 49–52.

Bello, D. C. (1992). Industrial buyer behavior at trade shows: Implications for selling effectiveness. *Journal of Business Research, 25*, 59–80.

Bello, D. C., & Barksdale, H. C. (1986). Exporting at industrial trade dhows. *Industrial Marketing Management, 15*, 197–206.

Benhabib, J., & Spiegel, M. M. (1994). The role of human capital in economic development: Evidence from aggregate cross-country data. *Journal of Monetary Economics, 34*, 143–173.

Black, R. (1986). *The trade show industry: Management and marketing career opportunities*. Trade Show Bureau.

Bloom, D. E., Sachs, J. D., Collier, P., & Udry, C. (1998). Geography, demography and economic growth in Africa. *Brookings Papers on Economic Activity, 2*, 207–295.

Bonoma, T. V. (1983). Get more out of your trade show. *Harvard Business Review* (January–February), 75–83.

Borghini, S., Golfetto, F., & Rinallo, D. (2004). Using anthropological methods to study industrial marketing and purchasing: An exploration of professional trade shows. In *Presented at the industrial marketing purchasing conference*. Retrieved on May 11, 2022, from http://www.impgroup.org/paper_view.php?viewPaper=4505.

Borghini, S., Golfetto, F., & Rinallo, D. (2006). Ongoing search among industrial buyers. *Journal of Business Research, 59*, 1151–1159.

Braudel, F. (1992). *The perspective of the world*. University of California Press.

Brinkley, I. (2006). *Defining the knowledge economy: Knowledge economy programme report*. The Work Foundation.

Brunetti, A. (1997). Political variables in cross-country growth analysis. *Journal of Economic Surveys, 11*(2), 163–190.

Camagni, R. (1991). Local 'milieu', uncertainty and innovation networks: Towards a new dynamic theory of economic space. In R. Camagni (Ed.), *Innovation networks: Spatial perspectives* (pp. 121–144). Belhaven.

Carreras, A., & Torra, L. (2005). *Why did modern trade fairs appear?* Working Paper 874. Department of Economics and Business, Universitat Pompeu Fabra. Retrieved on May 23, 2022, from http://www.ebha.org/ebha2005/pdf%20papers/Carreras_Torra.pdf.

CES. (2011, January 6–9). *International consumer electronics show*. Retrieved May 2011, from http://www.cesweb.org.

CMTS. (2009, October 19–22). *Canadian manufacturing technology show*. Post Show Report. Canadian Manufacturing Technology Show. Retrieved October 2010, from http://www.cmts.ca.

Colechia, A., & Schreyer, P. (2002). ICT investment and economic growth in the 1990s: Is the United States a unique case? A comparative study of nine OECD countries. *Review of Economic Dynamics, 5*(2), 408–442.

Cooke, P., & Morgan, K. (1998). *The associational economy: Firms, regions, and innovation*. Oxford University Press.

David, P. A., & Foray, D. (2002). An introduction to the economy of the knowledge society. *International Social Science Journal, 54*(171), 9–23.

De Long, J. B., & Summers, L. H. (1991). Equipment investment and economic growth. *The Quarterly Journal of Economics, 106*(2), 445–502.

Dekimpe, M., Francois, P., Gopalakrishna, S., Lilien, G. L., & Van den Bulte, C. (1997). Generalizing about trade show effectiveness: A cross-national comparison. *The Journal of Marketing, 61*(4), 55–64.

Donckels, R., & Lambrecht, J. (1995). Networks and small business growth: An explanatory model. *Small Business Economics, 7*(4), 273–289.

Dosi, G. (1995). Hierarchies, markets and power: Some foundational issues on the nature of contemporary economic organizations. *Industrial and Corporate Change, 4*(1), 1–19.

Drucker, P. F. (1998). The discipline of innovation. *Harvard Business Review, 76*(6), 149–157.

Easterly, W., & Levine, R. (2003). Tropics, germs and crops: How endowments influence economic development. *Journal of Monetary Economics, 50*(1), 3–39.

EMO. (2007, September 17–22). EMO Hannover 2007, Fair Profile. Retrieved December 2011, from http://www.emo-hannover.de/.

EMO. (2011, September 19–24). EMO Hannover 2011, Closing Press Release. Retrieved December 2011, from http://www.emo-hannover.de/.

Entwistle, J., & Rocamora, A. (2006). The field of fashion materialized: A study of London fashion week. *Sociology, 40*(4), 735–751.

Face, R. D. (1958). Techniques of business in the trade between the fairs of champagne and the south of Europe in the twelfth and thirteenth centuries. *The Economic History Review, 10*(3), 427–438.

Fagerberg, J. (1987). A technology gap approach to why growth rates differ. *Research Policy, 16*(2–4), 87–99.

Feiler, G. (1998). *From boycott to economic cooperation: The political economy of the Arab boycott in Israel*. Frank Cass Publishers.

Fischer, W. (1992). Zur Geschichte der Messen in Europa (On the history of trade fairs in Europe). In K.-H. Strothmann & M. Busche (Eds.), *Handbuch Messemarketing (Handbook of trade fair marketing)* (pp. 3–13). Gabler.

Flood, J. L. (2007). *'Omnium totius orbis emporiorum compendium'*: The Frankfurt fair in the early modern period. In R. Myers, M. Harris, & G. Mandelbrote (Eds.), *Fairs, markets and itinerant book trade* (pp. 1–40). Oak Knoll Press.

Frieden, J. A. (2006). *Global capitalism: Its fall and rise in the twentieth century*. W. W. Norton.

Fujita, M., Krugman, P. R., & Venables, A. J. (1999). *The spatial economy: Cities, regions, and international trade*. MIT Press.

Gallup, J. L., Sachs, J., & Mellinger, A. D. (1999). Geography and economic growth. *International Regional Science Review, 22*(2), 179–232.

Gibson, R., & Bathelt, H. (2014). Field configuration or field reproduction? *Zeitschrift für Wirtschaftsgeographie, 58*(1), 216–231.

Gilpin, R. (2001). *Global political economy: Understanding the international economic order*. Princeton University Press.

Godar, S. H., & O'Connor, P. J. (2001). Same time next year – Buyer trade show motives. *Industrial Marketing Management, 30*(1), 77–86.

Godin, B. (2006). The knowledge-based economy: Conceptual framework or buzz-word? *The Journal of Technology Transfer, 31*(1), 17–30.

Golfetto, F. (2004). *Fiere e communicazione: Strumenti per le impresse e il territorio (Trade fairs and communication: Tools for firms and the territory)*. EGEA.

Gopalakrishna, S., & Lilien, G. L. (1995). A three-stage model of industrial trade show performance. *Marketing Science, 14*(1), 22–42.

Granovetter, M., & Swedberg, R. (Eds.). (1992). *The sociology of economic life*. Westview Press.

Grier, K. B., & Tullock, G. (1989). An empirical analysis of cross-national economic growth, 1951–1980. *Journal of Monetary Economics, 24*(2), 259–276.

Hall, P. A., & Soskice, D. (2001). An introduction to varieties of capitalism. In P. A. Hall & D. Soskice (Eds.), *Varieties of capitalism: The institutional foundations of comparative advantage* (pp. 1–68). Oxford University Press.

Hansen, K. (2004). Measuring performance and trade shows: Scale development and validation. *Journal of Business Research, 57*, 1–13.

Hanushek, E. A., & Kimko, D. D. (2000). Schooling, labor-force quality, and the growth of nations. *American Economic Review, 90*, 1184–1200.

Hardy, C., & Maguire, S. (2010). Discourse, field-configuring events, and change in organizations and institutional fields: Narratives of DDT and the Stockholm convention. *Academy of Management Journal, 53*(6), 1365–1392.

Helmke, G., & Levitsky, S. (2004). Informal institutions and comparative politics: A research agenda. *Perspectives on Politics, 2*(4), 725–740.

Imp-Exp Executive Magazine (Sail by Exhibition). (2010). *Top 100 world trade fairs*. Retrieved June 2018, from http://www.prnewswire.com/news-releases/top-100-world-trade-fairs-2010-98308724.html.

Kerin, R. A., & Cron, W. L. (1987). Assessing trade show functions and performance: An exploratory study. *The Journal of Marketing, 51*(3), 87–94.

Knack, S., & Keefer, P. (1995). Institutions and economic performance: Cross-country tests using alternative institutional measures. *Economics and Politics, 7*(3), 207–227.

Koelble, T. A. (1995). The new institutionalism in political science and sociology. *Comparative Politics, 27*(2), 231–243.

Kormendi, R. C., & Meguire, P. G. (1985). Macroeconomic determinants of growth: Cross-country evidence. *Journal of Monetary Economics, 16*(4), 141–163.

Krugman, P. (1991). Increasing returns and economic geography. *Journal of Political Economy, 99*(3), 183–199.

Lam, A., & Lundvall, B.-Å. (2006). The learning organisation and national systems of competence building and innovation. In E. Lorenz & B.-Å. Lundvall (Eds.), *How Europe's economies learn: Coordinating competing models* (pp. 109–139). Oxford University Press.

Lampel, J., & Meyer, A. D. (2008). Guest editors' introduction: Field-configuring events as structuring mechanisms: How conferences, ceremonies, and trade shows constitute new technologies, industries, and markets. *Journal of Management Studies, 45*(6), 1025–1035.

Lee, D., & Hudson, D. (2004). The old and new significance of political economy in diplomacy. *Review of International Studies, 30*(3), 343–360.

Lichtenberg, F. R. (1992). *R&D investment and international productivity differences*. NBER Working Paper, No. 4161.

LightFair International. (2009a). *Official on-site directory*. Brochure.

LightFair International. (2009b). *General exhibit information*. Catalogue.

Lipset, S. M. (1959). Some social requisites of democracy: Economic development and political legitimacy. *The American Political Science Review, 53*(1), 69–105.

Lucas, R. E. (1988). On the mechanics of economic development. *Journal of Monetary Economics, 22*, 3–42.

Lundvall, B.-Å., & Foray, D. (1998). The knowledge-based economy: From the economics of knowledge to the learning economy. In D. Neef, G. A. Siesfeld, & J. Cefola (Eds.), *The economic impact of knowledge* (pp. 115–122). Butterworth-Heinemann.

Lundvall, B.-Å., & Johnson, B. (1994). The learning economy. *Journal of Industry Studies, 1*(2), 23–42.

Lundvall, B.-Å., & Johnson, B. (2016). The learning economy. In B.-Å. Lundvall (Ed.), *The learning economy and the economics of hope* (pp. 107–129). Anthem Press.

Mankiw, N. G., Romer, D., & Weil, D. N. (1992). A contribution to the empirics of economic growth. *Quarterly Journal of Economics, 107*(2), 407–437.

March, J. G., & Olsen, J. P. (1989). *Rediscovering institutions*. Free.

Maskell, P., & Malmberg, A. (1999a). The competitiveness of firms and regions: 'Ubiquitification' and the importance of localised learning. *European Urban and Regional Studies, 6*(1), 9–25.

Maskell, P., & Malmberg, A. (1999b). Localised learning and industrial competitiveness. *Cambridge Journal of Economics, 23*(2), 167–185.

Maskell, P., Bathelt, H., & Malmberg, A. (2006). Building global knowledge pipelines: The role of temporary clusters. *European Planning Studies, 14*, 997–1013.

Masters, W. A., & McMillan, M. S. (2001). Climate and scale in economic growth. *Journal of Economic Growth, 6*(3), 167–186.

Mauro, P. (1995). Corruption and growth. *Quarterly Journal of Economics, 110*(3), 681–712.

Messe Mūnchen GmbH. (2010a, November 2010). *Electronica: Electronica 2010 confirms upward trend in the electronics industry*. Final Report. Messe Mūnchen GmbH.

Messe Mūnchen GmbH. (2010b). *Electronica 2010: A record of success*. Review 2010. Messe Mūnchen GmbH.

Meyer, A. D., Gaba, V., & Colwell, K. (2005). Organizing far from equilibrium: Non-linear change in organizational fields. *Organization Science, 16*, 456–473.

Milgrom, P. R., North, D. C., & Weingast, B. R. (1990). The role of institutions in the revival of trade: The law merchant, private judges, and the champagne fairs. *Economics and Politics, 2*(1), 1–23.

MMTS. (2010). *Montreal manufacturing technology show*. Post Show Report. Retrieved October 2010, from http://www.mmts.ca.

Moeran, B. (2011). Trade fairs, markets and fields: Framing imagined as real communities. *Historical Social Research, 36*(3), 79–98.

Moeran, B., & Pedersen, J. S. (Eds.). (2011). *Negotiating values in the creative industries: Fairs, festivals and competitive events*. Cambridge University Press.

Munro, J. H. (2001). The 'new institutional economics' and the changing fortunes of fairs in medieval and early modern Europe: The textile trades, warfare, and transaction costs. *VSWG – Vierteljahrschrift für Sozial – und Wirtschaftsgeschichte, 88*(1), 1–47.

Myrdal, G. (1957). *Economic theory and underdeveloped regions*. Hutchinson Publications.

North, D. C. (1990). *Institutions, institutional change and economic performance*. Cambridge University Press.

North, D. C., & Thomas, R. P. (1973). *The rise of the western world: A new economic history*. Cambridge University Press.

NPE. (2009, June 22–26). *International plastics exhibition show review*. Retrieved July 2009, from http://www.npe.org.

Nyqvist, A., Leivestad, H. H., & Tunestad, H. (2017). Individuals and industries: Large-scale professional gatherings as ethnographic fields. In H. H. Leivestad & A. Nyqvist (Eds.), *Ethnographies of conferences and trade fairs: Shaping industries and creating professionals* (pp. 1–21). Palgrave Macmillan.

Oatley, T. (2016). *International political economy*. Taylor & Francis.
O'Brien, R., & Williams, M. (2016). *Global political economy: Evolution and dynamics* (5th ed.). Palgrave Macmillan.
Oliver, A. L., & Montgomery, K. (2008). Using field-configuring events for sense-making: A cognitive network approach. *Journal of Management Studies, 45*(6), 1147–1167.
Organisation for Economic Co-operation and Development. (1996). *The knowledge-based economy*. OECD Publishing.
Organisation for Economic Co-operation and Development. (1999). *Managing national innovation systems*. OECD Publishing.
Petrakos, G., & Arvanitidis, P. (2008). Determinants of economic growth. *Economic Alternatives, 1*, 49–69.
Podrecca, E., & Carmeci, G. (2001). Fixed investment and economic growth: New results on causality. *Applied Economics, 33*, 177–182.
Polanyi, M. (1967). *The tacit dimension*. Anchor Books.
Porter, M. E. (1990). *The competitive advantage of nations*. McMillan Press Ltd..
Powell, W. W., & DiMaggio, P. J. (Eds.). (1991). *The new institutionalism in organizational analysis*. University of Chicago Press.
Pritchett, L. (2001). Where has all the education gone? *World Bank Economic Review, 15*, 367–391.
Prüser, S. (2003). Die messe als networking-plattform (Trade shows as a platform for networking). In M. Kirchgeorge, W. M. Dornscheidt, W. Giese, & N. Stoeck (Eds.), *Handbook of trade show management: Planning, execution and control of trade shows, conventions and events* (pp. 210–270). Gabler.
Ravenhill, J. (2008). *Global political economy* (2nd ed.). Oxford University Press.
Reid, D. A., & Plank, R. E. (2004). *Fundamentals of business marketing research*. The Haworth Press.
Rinallo, D., & Golfetto, F. (2011). Exploring the knowledge strategies of temporary cluster organizers: A longitudinal study of the EU fabric industry trade shows (1986–2006). *Economic Geography, 87*(4), 453–476.
Rinallo, D., Golfetto, F., & Borghini, S. (2007). The influence of occupational communities on buying behavior. In *Paper presented at the industrial marketing and purchasing conference*.
Roberts, J. (2000). From know-how to show-how? Questioning the role of information and communication technologies in knowledge transfer. *Technology Analysis & Strategic Management, 12*(4), 429–443.
Rodekamp, V. (2003). Zur Geschichte der Messen in Deutschland und Europe (On the history of trade fairs in Germany and Europe). In M. Kirchgeorge, W. M. Dornscheidt, W. Giese, & N. Stoeck (Eds.), *Handbuch Messemanagement: Planung, Durchführung und Kontroll von Messen, Kngressen und events (Handbook of trade show management: Planning, execution and control of trade shows, conventions and events)* (pp. 5–13). Gabler.
Rodrik, D. (1999). Where did all the growth go? External shocks, social conflict and growth collapses. *Journal of Economic Growth, 4*(4), 385–412.
Rodrik, D., (2003). *In search of prosperity: analytical narratives on economic growth*. Princeton University Press.
Rodrik, D., Subramanian, A., & Trebbi, F. (2002). Institutions rule: The primacy of institutions over geography and integration in economic development. *Journal of Economic Growth, 9*(2), 131–165.
Romer, P. (1986). Increasing returns and long run growth. *Journal of Political Economy, 94*(2), 1002–1037.
Rosson, P. J., & Seringhaus, F. H. R. (1995). Visitor and exhibitor interaction at industrial trade fairs. *Journal of Business Research, 32*(1), 81–90.
Ruggie, J. G. (1993). Territoriality and beyond: Problematizing modernity in international relations. *International Organization, 47*(1), 139–174.
Sachs, J., & Warner, A. (1997). Sources of slow growth in African economies. *Journal of African Economies, 6*(3), 335–376.

Sala-i-Martin, X. X. (1996). The classical approach to convergence analysis. *The Economic Journal, 106*(437), 1019–1036.

Sashi, C. M., & Perretty, J. (1992). Do trade shows provide value? *Industrial Marketing Management, 21*(3), 249–255.

Schuldt, N., & Bathelt, H. (2011). International trade fairs and global buzz. Part II: Practices of global buzz. *European Planning Studies, 19*(1), 1–22.

Schüssler, E., Rüling, C.-C., & Wittneben, B. (2014). On melting summits: The limitations of field-configuring events as catalysts of change in transnational climate policy. *Academy of Management Journal, 57*(1), 140–171.

Scully, G. (1988). The institutional framework and economic development. *Journal of Political Economy, 96*(3), 652–662.

Seringhaus, R., & Rosson, P. (2004). An analysis model for performance measurement of international trade fair exhibitors. *Problems and Perspectives in Management, 4*, 152–165.

Sharland, A., & Balogh, P. (1996). The value of nonselling activities at international trade shows. *Industrial Marketing Management, 25*(1), 59–66.

Shoham, A. (1992). Selecting and evaluating trade shows. *Industrial Marketing Management, 21*(4), 335–341.

Shoham, A. (1999). Bounded rationality, planning, standardization of international strategy, and export performance: A structural model examination. *Journal of International Marketing, 7*(2), 24–50.

Short, J., Williams, E., & Christie, B. (1976). *The social psychology of telecommunications*. Wiley.

Snowdon, B., (2003). *Conversations on growth, stability and trade: An historical perspective*. Edward Elgar.

Soete, L. L. (2006). *A knowledge economy paradigm and its consequences*. Merit Working Papers 001, United Nations University- Maastricht Economic and Social Research Institute on Innovation and Technology (MERIT).

Solow, R. (1956). A contribution to the theory of economic growth. *Quarterly Journal of Economics, 70*, 65–94.

Soludo, C., Ogbu, O., & Chang, H.-J. (2004). *The politics of trade and industrial policy in Africa: Forced consensus?* Africa World Press.

Steinmo, S., Thelen, K., & Longstreth, F. (Eds.). (1992). *Structuring politics: Historical institutionalism in comparative analysis*. Cambridge University Press.

Taylor, M. Z., (2004). Empirical evidence against Varieties of Capitalism's theory of technological innovation. *International Organization, 58*(3), 601–631.

Ulku, H. (2004). *R&D, Innovation and economic growth: An empirical analysis*. International Monetary Fund, IMF Working Paper 185.

van Hemert, P., & Nijkamp, P. (2011). Critical success factors for a knowledge-based economy: An empirical study into background factors of economic dynamism. In P. Nijkamp & I. Siedschlag (Eds.), *Innovation, growth and competitiveness: Dynamic regions in the knowledge-based world economy* (pp. 61–89). Springer.

Verlinden, C. (1963). Markets and fairs. In M. M. Postan, E. E. Rich, & E. Miller (Eds.), *The Cambridge economic history of Europe, Vol. III. Economic organization and policies in the middle ages* (pp. 119–153). Cambridge University Press.

Weber, A. (1909). *Über den standort der industrien* (Theory of the location of industries). J.C.B. Mohr.

Weisgal, M. B. (1998). *The end of the show*. Center for Exhibition Industry Research (CEIR), Report SM35.

Chapter 3
The Study of Comparative Capitalisms: The Problem of Institutional Change

3.1 Introduction

Having provided some insight into the role of international trade fairs in enabling knowledge flows and interaction processes between firms from different parts of the world, this chapter directs attention to the overarching institutional foundations upon which such inter-firm interactions are based. Responding to calls (see, for example, Deeg & Jackson, 2007) to engage multiple levels in the analysis of dynamic capitalist systems, this chapter proceeds to set the macro-level foundations of the theoretical discussion. To foreshadow core arguments, I suggest that the specific collection of institutions associated with different models of capitalism more or less strongly guide the actions and choices of firms that participate in international trade fairs. In keeping with the broader literature comparing capitalist systems, and, perhaps most notably, the VoC approach, this chapter suggests that these institutional configurations typically encourage firms to search for new technological ideas, solutions, and business partners that closely correspond to the existing or established production models found in their home countries. Such a finding supports the basic assumptions of the VoC approach and the comparative capitalisms (CC) literature more broadly. What will become clear, however, when the study's focus shifts to micro-level processes in Chap. 4, is that these technological search processes, while guided by the broader institutional frameworks of different capitalist varieties, rarely follow a rational or linear pattern. Instead, the search processes of firms at trade fairs are better captured through the concept of "organized anarchies" and the garbage-can processes originally conceived by Cohen et al. (1972) and further developed by others, notably Kingdon (2003).

The remainder of this chapter will be structured as follows. It begins by exploring the nature of national patterns of technological specialization in the context of the global political economy. Drawing on a variety of literatures in, for instance, economics and regional studies, it becomes clear that despite the speed and

R. Gibson, *International Trade Fairs and Inter-Firm Knowledge Flows*,
https://doi.org/10.1007/978-3-031-20557-6_3

pervasiveness of globalization and internationalization processes, many countries continue to specialize in different sectors of the economy, revealing specific "sectors of excellence" at the level of the national state (Archibugi & Pianta, 1992). With this empirical puzzle in mind, the chapter highlights the duality of leading international trade fairs. More specifically, I suggest that such events could, on the one hand, operate and be used as critical instruments for enabling and accelerating processes of economic globalization. Yet, on the other hand, leading international trade fairs could also be a decisive mechanism for strengthening existing patterns of national technological specialization. These different trajectories and their underlying conditions will be explored in greater detail at a later point in the study.

Next, I provide a stylized review and synthesis of the comparative capitalisms (CC) literature (Jackson & Deeg, 2006; Deeg & Jackson, 2007), which, in the present formulation, includes four analytical approaches in particular: Varieties of Capitalism (VoC); National Systems of Innovation (NSI); National Business Systems (NBS); and Economic Governance (EG) or simply Governance approaches. This literature review draws attention to the unifying features among the different approaches and follows Jackson and Deeg (2006; 2008; see also Deeg & Jackson, 2007) in suggesting that, despite some conceptual and methodological differences, there is sufficient common ground among the four approaches to identify one general CC approach. This section of the chapter also addresses the main limitations of the CC literature, drawing particular attention to two recent critiques that serve as the basis for the current study.

Focusing on one of the CC literature's core limitations, the third section of the chapter engages with the problem of institutional change within the context of comparative politics and the related "institutionalist" literature in other social science fields. In particular, I examine recent efforts to theorize more gradual forms of institutional change, as opposed to the more traditional focus on periods of abrupt, wide-ranging institutional transformation (Thelen, 2004; Mahoney & Thelen, 2010). The chapter concludes by summarizing the main points raised and by introducing Chap. 4, which represents the final major theoretical contribution to the study's overarching analytical framework.

3.2 National Patterns of Technological Specialization

In general terms, technological specialization may be defined as "the structure of the distribution of activities across different sectors of a country" (Archibugi & Pianta, 1992, 1). When speaking of national patterns of technological specialization, it is important to note that the technological capabilities of any given country typically involve a diverse range of knowledge and innovation sources, including activities that are easily codified as well as those that are more tacit in nature (Lundvall, 1992; Archibugi & Coco, 2004). Some technological capabilities may be disembodied in character, as seen in the form of new ideas or innovations (Archibugi & Coco, 2004), whereas others may be embodied in, for example, equipment, machines, and human capital (Pianta, 1995; Evangelista, 1999; Archibugi & Coco, 2004).

Distinct national patterns of industry and technological specialization are evident in many countries throughout the world. In their detailed report on the subject of technological specialization in advanced countries, Archibugi and Pianta (1992) describe the specialization patterns of different national states using data from different patenting institutions. They find, for instance, that Japan's most significant technological advantages reside in electrical and electronic areas, including electronics, telecommunications, optics, photographic equipment, and computing. More recent studies confirm these findings, suggesting that Japan continues to enjoy competitive industrial strengths in the area of complex assembly manufacturing. Haake's (2002) research, for example, highlights the significance of electronics, in particular, but also points to cameras, machine tools, and cars as other specific examples of technological advantage. On the other hand, Japan reveals weaknesses in the more traditional industries like foodstuffs, footwear, clothing and textiles (Archibugi & Pianta, 1992).

Other countries reveal distinct patterns of specialization as well. The United States, for example, demonstrates particular strengths in the high-technology industries, as well as in a wide range of services industries, from financial services to film production (Haake, 2002). Both in terms of domestic and foreign markets, the U.S. tends to be relatively weak in areas such as textiles, machine tools, and transportation (Archibugi & Pianta, 1992). In the German setting, chemicals represent the strongest field of technological advantage. Germany's specialization profile also includes other manufacturing industries, such as weapons, building, transport, hand tools, engines, and luxury vehicles (Archibugi & Pianta, 1992). A notable de-specialization is seen in Germany in electrical areas such as electronics and telecommunications (Archibugi & Pianta, 1992).

The persistence of national patterns of technological specialization in the face of ongoing processes of globalization represents an interesting and important puzzle for scholars interested in the role of international trade fairs in the global political economy. On the one hand, the activities of firms at leading international events closely correspond to an aspect of economic globalization that has been referred to as "ubiquitification" (Maskell & Malmberg, 1999). As mentioned in Chap. 2, the term "ubiquitification" denotes the process by which previously localized firm capabilities are converted into ubiquities that are available at approximately the same cost throughout the world. Driven largely by processes associated with globalization, including the increasing codification and dissemination of technical knowledge on a global scale, ubiquitification directly challenges the continuation of distinct capitalist varieties by empowering processes of convergence across such political economies with respect to the technological specialization of firms. Maskell and Malmberg (1999, 12) describe the problem as follows:

> No firm can build competitiveness on ubiquities alone, and little economic progress would be made anywhere if everyone were able to do exactly the same in all places at once. In order to enhance the competitiveness of firms, a localized capability must thus be valuable, and in order to be valuable it has to be rare. Without being rare, there is no way in which a

> regional or national capability can be valuable. If, however, a formerly important and rare localized capability is somehow turned into a ubiquity – making the localized capability equally available at the same cost to all firms more or less regardless of location – the capability loses its importance. Firms whose competitiveness depended on it will be penalized in the market just as, at an aggregate level, the established patterns of regional or national specialization will be jeopardized.

As the above quote suggests, the ramifications of ubiquitification – a process which trade fairs may actively contribute to – are serious: "as ubiquities are created, localized capabilities are destroyed" (Maskell & Malmberg, 1999, 12). Yet, as previously discussed, many countries continue to demonstrate decisive competitive advantages in particular technologies and industries (Scott & Lodge, 1985; Porter, 1990; Patel & Pavitt, 1991; Nelson, 1993; Hollingsworth et al., 1994). Whereas patterns of continued national specialization are generally supported by the CC literature, such patterns are puzzling in light of conventional understandings of the role of international trade fairs in global economic processes. This study aims to shed some light on this empirical puzzle.

3.3 The Study of Comparative Capitalisms: A Review of the Literature

Over the last 25 years or so, a substantial and eclectic literature has developed based on the notion of distinct capitalist varieties. This literature exhibits a great deal of conceptual and methodological diversity, and is also highly interdisciplinary in nature, drawing on a wide range of institutional perspectives from different academic fields and subfields, including political science, economic sociology, and institutional economics (Sorge & Warner, 1986; Aoki, 2001; Hall & Soskice, 2001). Although the literature comparing different capitalist systems is indebted to a long and disciplinarily broad intellectual tradition that includes, for example, Andrew Shonfield (1965), Alfred Chandler (1978), and Michael Porter (1990),[1] this chapter concentrates exclusively on four main analytical approaches to the comparative study of capitalist diversity. As mentioned above, these are the VoC approach, the NSI perspective, the NBS approach, and EG approaches.[2] Despite the lack of a unified intellectual approach, these individual literatures have contributed a great deal to our understanding of capitalist diversity and they share several unifying

[1] For a more in-depth discussion of the historical development of the "comparative capitalisms" literature, see Jackson and Deeg (2006, 7–11); and Jackson and Deeg (2008, 681–684).

[2] In defining the comparative capitalisms (CC) literature, Jackson and Deeg (2006) (see also, Deeg & Jackson, 2007; Jackson & Deeg, 2008) only include three of these main analytical perspectives: the Varieties of Capitalism approach, Governance approaches, and the National Business Systems approach. While the present literature review closely follows the work of these authors, it argues that the National Systems of Innovation approach represents another important analytical perspective that fits within the broader CC tradition, according to the defining criteria laid out by Jackson and Deeg (2006).

characteristics that have led scholars to recently distinguish a broader comparative capitalisms literature or tradition (Jackson & Deeg, 2006, 2008; Deeg & Jackson, 2007).

The present literature review highlights the common features among these different bodies of work and identifies specific attributes that set the CC literature apart from other related literatures, such as the "new comparative economics" (Djankov et al., 2003). The literature review also discusses the strengths and limitations of the CC literature, specifically as they relate to questions of institutional change. In doing so, this section of the chapter closely follows the work of Jackson and Deeg (2006; see also, Deeg & Jackson, 2007), and ultimately suggests that there is a need for more dynamic approaches to the study of capitalist diversity. However, in this study, I aim to go one step further by developing an empirically-grounded analysis that responds to this specified weakness in the literature – that is, the inability of existing approaches to adequately capture the dynamic aspects of capitalist variation. While the CC literature's general preoccupation with a broad notion of institutional path dependence may well be the source of this problem, this study suggests that new insights regarding incremental processes of institutional change may help overcome such barriers without disregarding the persuasive power of historical legacies.

In this book, my aim is to inject greater dynamism into the comparative study of capitalist economies by introducing a new, and still largely under-examined, source of economic dynamism – namely, trade fairs. I argue that trade fairs should be viewed as a crucial meso-level structure that creates a closer linkage between broad macro-level patterns (in this case, the potential future structure of national patterns of technological specialization) and micro-level phenomena (in this case, the technological search and information processes of exhibiting firms at trade fairs). But before moving on to develop this argument, it is necessary to first establish the macro foundations of the research. I do this first by providing an overview of what has arguably become the most influential analytical framework in the study of capitalist varieties over the last 15 years: the VoC approach.

3.3.1 Varieties of Capitalism (VoC)

Since the publication of Hall and Soskice's seminal volume, *Varieties of Capitalism: The Institutional Foundations of Comparative Advantage* (2001), the VoC approach has become nearly synonymous with the comparative institutional analysis of capitalist systems. As Howell (2003, 103) observes, Hall and Soskice's (2001) VoC approach has become "close to hegemonic in the field" of political economy. The VoC approach is based on two key conceptual foundations: "institutional complementarities" and "system coordination" (Hancké et al., 2007). Taken together, these two notions form the basis of an approach that argues that a country's particular institutional make-up plays a decisive role in determining the nature and direction of its economic development and technological specialization patterns. Furthermore, it suggests that these unique institutional configurations allow for the identification and comparison of different capitalist varieties based on a continuum model of firms' coordination activities (Hall & Soskice, 2001).

In the context of the broader CC literature, the original VoC approach stands apart from other related perspectives in that it proposes a decidedly "firm-centred" approach based on a relational conception of the firm. In other words, the VoC approach draws explicit attention to the complex actions and interactions of economic agents operating within diverse national contexts (Hall & Soskice, 2001). Whereas the VoC approach finds common ground with other related perspectives in the sense that it aims to better understand broad institutional similarities and differences among capitalist economies (Hall & Soskice, 2001), it departs from most other institutionalist theories in its conceptualization of institutions. From the VoC perspective, the significance of institutions relates specifically to their role in organizing economic interactions and addressing the coordination problems of firms within different spheres of economic activity (Soskice, 1999; Howell, 2003). The central claim here is that institutions are responsible for much more than simply distributing power and sanctioning behaviour; they also play a critical role in enabling processes of ongoing deliberation and knowledge exchange among economic actors. Although this view of institutions finds important parallels with recent conceptualizations of international trade fairs as critical platforms for knowledge creation and exchange within the context of the global knowledge economy, VoC scholars have not yet acknowledged these events as a potential area of inquiry.

According to Hall and Soskice (2001), there are five spheres in which firms must establish relationships in order to solve coordination problems pertinent to their core capabilities. These five spheres are: industrial relations, vocational training and education, corporate governance, inter-firm relations, and employees.[3] While I recognize the interdependent nature of these separate spheres, the present study focuses specifically on the domain of inter-firm relations, which denotes the relationships a firm creates with other companies, particularly suppliers and customers. Such relationships are crucial in building or maintaining steady demand for a firm's products or services, finding suitable suppliers of inputs, and ensuring access to new technologies and innovations (Hall & Soskice, 2001). Achieving such ends may require firms to engage in joint standard-setting practices, develop mechanisms to share technological knowledge, and find appropriate firms to partner with in research and development activities (Hall & Soskice, 2001). Establishing such suitable and durable relationships in the sphere of inter-firm relations depends largely on "the capacities of firms to remain competitive and technological progress in the economy as a whole" (Hall & Soskice, 2001, 7). I argue that leading international trade fairs can play an important role in the formation of such inter-firm relationships and,

[3]For a more detailed description of each of the five spheres, see Hall and Soskice (2001, 6–7).

consequently, in finding potential resolutions to complex coordination problems within this institutional domain.

The coordination of economic activity differs in ways that produce observable patterns between capitalist systems. These distinct national systems of interconnected institutions are largely resistant to ongoing pressures for change brought about, for instance, by the various forces associated with economic globalization and internationalization. The VoC approach aims to explain these distinctive patterns through a theory of "comparative institutional advantage", whereby firms are rewarded for participating in specific kinds of activities that correspond to the institutional structure or "institutional complementarities" of the national system in which they are embedded (Soskice, 1999; Hall & Soskice, 2001).[4] Indeed, the performance or successful function of any given institution depends largely on the accompaniment of other institutions, with institutions having a propensity to become deeply interconnected and reinforce one another across different domains of economic activity over time (Howell, 2003). In Hall and Soskice's (2001) formulation, such institutional complementarities have facilitated the placement of most western capitalist countries in one of two ideal-typical forms of economic organization, each possessing a unique assortment of complementary institutions that addresses the coordination challenges of firms in fundamentally different ways. These two ideal-types are: liberal market economies (henceforth referred to as LMEs), of which the classic example is the United States, and coordinated market economies (henceforth CMEs), which are exemplified by the German case.

In LMEs, firms rely mainly on hierarchies and competitive market relationships in order to coordinate their economic activities and build core competencies. In CMEs, on the other hand, firms depend more extensively on non-market forms of coordination for such undertakings (Hall & Soskice, 2001). Market-based modes of coordination are defined by distant and independent exchanges of goods and services in a competitive setting that relies primarily on formal contracting; non-market modes of coordination usually involve "more extensive relational or incomplete contracting, network monitoring based on the exchange of private information inside networks, and more reliance on collaborative [...] relationships to build the competencies of the firm" (Hall & Soskice, 2001, 8). In CMEs, strategic interaction among firms and other economic agents, thus, plays a pivotal role in determining economic outcomes, whereas the dynamics

[4] In terms of its implications regarding the persistence of different capitalist systems, the VoC literature draws conclusions similar to those in regulation theory (Lipietz, 1987; Boyer, 1990), the social-systems-of-production approach (Hollingsworth & Boyer, 1997), or the national-business-systems literature (Whitley, 1999). Due to its relational perspective, however, the VoC approach directly allows us to integrate a micro-level analysis with macro-level patterns.

of supply and demand conditions are of central importance in LMEs (Hall & Soskice, 2001).[5]

The VoC approach makes a number of critical contributions to the comparative analysis of capitalist diversity. For instance, it uses a micro-economic perspective that has proven extremely valuable in elucidating broad economic processes and trajectories at the level of the national state. In more specific terms, the VoC approach relies on a relational perspective, which involves looking at the actions and interactions of agents such as firms, other organizations, and the managers and employees that operate within them, in order to better understand and explain complex macro-economic patterns. Such patterns may include national patterns of technological specialization, which are the primary interest in this study. Beyond this, the VoC approach develops a useful methodological framework for analyzing and systematically comparing the various ways in which firms from different capitalist systems coordinate their economic activities. This is achieved through the elaboration of five institutional spheres (mentioned above) within which economic action takes place.

As initially conceived, the VoC approach makes a deliberate attempt to distinguish itself from other related literatures, such as the NSI literature (Hall & Soskice, 2001, 3 f.). While it is true that early classifications of national innovation systems often support the VoC's dichotomous distinction between LMEs and CMEs (Lorenz & Lundvall, 2006, 13–14), more recent studies have offered increasingly nuanced and sophisticated accounts of the different types of knowledge and learning processes that take place between firms from different national economic systems (Lorenz & Lundvall, 2006). Despite these recent developments, the two literatures share several important properties, such as a view of economic action being socially-embedded, an appreciation of the interconnected nature of institutions, and general adherence to a broad notion of path dependence.

Beyond these similarities, the VoC approach corresponds to the NSI approach in suggesting that innovation represents a crucial type of comparative advantage for firms, and by directly linking institutions to processes of innovation. To briefly elaborate, the VoC approach distinguishes between radical and incremental forms of innovation, suggesting that LMEs enjoy a comparative advantage in radical innovations – that is, innovations that involve abrupt shifts in product lines, substantial changes in production processes, or the introduction of brand-new goods or services (Hall & Soskice, 2001). CMEs, on the other hand, have a comparative advantage in incremental innovations, or innovations that entail continuous but more subtle shifts

[5] In their original formulation of the approach, Hall and Soskice (2001) clearly acknowledge that market relationships and hierarchies are significant for firms in all types of capitalist models, and that not all LMEs are perfectly competitive. In their view, the LME/CME dichotomy is rooted in the assertion that "the incidence of different types of firm relationships varies systematically across nations" and that firms in a given national state will be drawn towards the particular mode of coordination that is institutionally supported in their home context (Hall & Soskice, 2001, 8–9).

in existing product lines or production processes (Hall & Soskice, 2001).[6] With these key points in mind, I now turn to what might be considered a second major analytical framework within the CC literature – namely, the national systems of innovation approach.

3.3.2 National Systems of Innovation (NSI)

In the broader CC literature, institutions are widely recognized as having an important role in shaping the innovative capacity of firms (Edquist, 1997; Whitley, 1999; Hall & Soskice, 2001; Werle, 2012). But the relationship between institutions and innovation takes centre stage in studies associated with the NSI approach. The systematic study of national innovation systems can be traced back to the 1980s and the scholarly work of Freeman (1982, 1987), Lundvall (1985), Nelson (1988), Lundvall and Freeman (1988), and Dosi (1988). A national system of innovation can be defined as:

> [A] set of distinct institutions which jointly and individually contribute to the development and diffusion of new technologies and which provides the framework within which governments form and implement policies to influence the innovation process. As such it is a system of interconnected institutions to create, store and transfer knowledge, skills and artifacts, which define new technologies. The element of nationality follows not only from the domain of technology policy but also from elements of shared language and culture which bind the system together, and from the national focus of other policies, laws and regulations which condition the innovative environment. (Metcalfe, 1997, 289)

As the above passage indicates, the concept of the national innovation system shares much in common with the basic precepts underlying the VoC approach. The notion of institutional interdependence or complementarities, for instance, is clearly

[6] Recent empirical work (Taylor, 2004, 2016; Akkermans et al., 2009; Schneider & Paunescu, 2012) has challenged the VoC's distinction between radical and incremental forms of innovation. Taylor (2004, 2016), for instance, suggests that linking different kinds of technological innovation to specific sectors neglects the fact that individual sectors can include both radical and incremental forms of technological innovation. Furthermore, Taylor (2004) also finds that the U.S. case, which is generally viewed as the classic example of a liberal market economy, constitutes an outlier in the data; when the U.S. case is removed from the analysis, the LME-CME distinction in relation to patterns of radical and incremental innovation becomes largely unsustainable. The study of Akkermans et al. (2009), which extends Taylor's (2004) work, finds more mixed results, suggesting that there is some but not full support for the VoC's assignment of comparative technological advantages to specific sectors. The present study is unable to speak directly to this debate, as the data collected do not distinguish between radical and incremental forms of innovation. Developing a more fine-grained analysis of the radical-incremental innovation dichotomy would be helpful in future research in this area. Yet, the focus of this book is on broader categories of core competencies (e.g., innovation focus, customized production, strong R&D focus) with the objective of developing a deeper understanding of how the behaviour of firms at international trade fairs facilitates or disrupts processes of divergence/convergence regarding such technological specialization patterns.

implied in this conceptualization. Moreover, the above definition nicely captures the idea that economic action is deeply rooted in a social context that provides specific "elements of a shared language and culture", to re-employ Metcalfe's (1997) words. Finally, the above passage strongly suggests that these socially-embedded institutions bring about a particular logic of economic action.

Focusing their analytical lens specifically on the dynamics of technological change, proponents of the NSI approach suggest that feedback processes between national production structures, institutions, and innovation patterns encourage the establishment of particular innovation "systems" at the national level (Lundvall, 1992; Nelson, 1993; Edquist, 1997). Distinct national patterns of innovation develop as existing specializations pre-structure the types of problems and bottlenecks in production that are prioritized in a particular national environment (Lundvall & Maskell, 2000). And this, in turn, leads to the formation of different national industrial systems and an institutional framework that encourages particular types of economic interaction, both of which strongly shape the future direction of innovation processes (Archibugi et al., 1999). From this perspective, such patterns of economic interaction depend on a wide range of factors, including the division of labour within and between firms, existing technological competencies in the workforce, the creation and reproduction of sophisticated skill sets, as well as other related aspects of the capital-labour nexus (Gertler, 2004).

The interdependence among patterns of production, institutional arrangements, and knowledge creation suggests that actors involved in innovation processes will enjoy particular comparative advantages when working with partners from within, rather than outside, their national context (Lundvall & Maskell, 2000). The nature of this interdependence is nicely captured in the following statement by Pavitt and Patel (1999, 103): "the technological competitiveness of firms inevitably depends on national systems of innovation, and national systems of innovation inevitably depend on government policy. The level of business-funded R&D is influenced by national policies (e.g., competition, macroeconomics), and also by the behaviour of national institutions (e.g., agencies funding basic research, banks and stock markets, systems of corporate governance)." As indicated above, institutional complementarities are further promoted by the presence of a common framework of understanding, a shared "language" so to speak, an intimate knowledge of the specifics of the technologies being used, and similar experiences in problem solving. In other words, economic action is strongly influenced by the social context in which it takes place.

Compared to the VoC approach, and the CC literature in general, the NSI approach draws greater attention to the particular spatial dynamics of innovation processes, documenting, for instance, how the behaviour of economic agents shapes and is shaped by the particular features of these spatial configurations. Care is taken not only to precisely define the components, but also the boundaries of these national systems of innovation. Yet, with regards to broader questions of institutional change and stability, the two approaches find much common ground. Despite the growing internationalization of economic and innovative activity, there is widespread agreement in the NSI literature that national innovation systems remain essential in order to generate and maintain firms' technological advantage on the global stage (Nelson, 1993; Freeman, 1995; Patel, 1997; Pavitt, 1998).[7] Differently expressed, national innovation systems have retained their significance in the face of increasing processes of economic globalization and internationalization: "far from being irrelevant, what happens in home countries is still very important in the creation of global technological advantage for firms" (Patel, 1997, 212). The continuing importance of these national innovation systems relates directly to the networks of relationships that underpin them (Freeman, 1995). Although it has become increasingly important for firms to develop external connections with economic actors outside of their national context (Bathelt & Schuldt, 2008), the influence of national-level institutions – such as industrial relations, technical and scientific institutions, government policies, and cultural norms and customs – remains instrumental (Freeman, 1995). The nature of these particular sets of institutions is what defines a given national innovation system. Moreover, because these institutions function interdependently, they have a tendency to evolve very slowly or incrementally, creating enduring patterns of national-level path dependence (Nelson, 1993).

As discussed in Chap. 2, trade fairs can play a pivotal role in processes of innovation and knowledge creation. Recent empirical work demonstrates, for instance, that the majority of trade fair exhibitors systematically use these events to display new innovations and obtain valuable information about the latest products and technologies (Bathelt et al., 2014). As such, trade fairs may be seen as an important part of the respective institutional configuration that supports the national innovation system as a whole. While temporary, these events may support existing institutional complementarities within a specific NSI by providing a common framework for understanding, interactive learning, and knowledge sharing around new technologies, for example, or solutions.

Out of necessity, the above overview has glossed over some of the important debates and developments in the NSI literature over the past 25 years; however, the primary aim has been to demonstrate that the NSI approach shares at least three major propositions that, according to Jackson and Deeg (2006; Deeg & Jackson, 2007), define the CC literature. To summarize, the NSI literature more or less explicitly draws on the concept of social "embeddedness" (Granovetter, 1985) when describing the complex dynamics of economic action; it conceptualizes the various institutions within the national economy as complementary or interdependent, thereby resulting in distinct national patterns or "systems" of

[7] It is important to note that some of this work includes the analysis of sub-national innovation systems as well, such as those operating at the regional or sectoral levels. This does not, however, take away from the basic argument presented above regarding the tendency of these approaches to adhere to a broad notion of path dependence.

innovation; and it suggests that, taken together, these two characteristics typically serve as strong impediments to change, resulting in the persistence of distinct national systems of innovation, rather than convergence on a single model (Nelson, 1993). As such, I suggest that the NSI approach can be appropriately situated within the broader CC tradition. In continuing to sketch out this wider CC tradition, the chapter now turns to another key representative of the literature: the national business systems (NBS) approach.

3.3.3 National Business Systems (NBS)

As indicated above, the conceptual development of the VoC approach runs parallel to theoretical developments in other related literatures. A third major strand of the CC literature is the national business systems (NBS) approach, which is associated, most notably, with the work of Richard Whitley (1999). The NBS approach is based on the assumption that different types of capitalism institutionalize different "rules of the game", which, in turn, play a decisive role in the organization, strategic behaviour, and competitiveness of firms operating within such national contexts (Whitley, 1999; Morgan, 2007). Drawing from interesting research on East Asian capitalism (e.g., Johnson, 1982; Biggart, 1991; Whitley, 1992), Whitley develops a systematic framework to distinguish and compare national "business systems", which he defines as "distinctive patterns of economic organization that vary in their degree and mode of authoritative coordination of economic activities, and in the organization of, and interconnections between, owners, managers, experts, and other employees" (Whitley, 1999, 33).

In Whitley's framework, there is less focus on the individual agency of firms and more emphasis on interactions between stakeholders, and on how such interactions are linked to the development of different firm capabilities (Whitley, 1999; Morgan, 2007; Jackson & Deeg, 2008). From this perspective, Whitley identifies two key factors that influence different types of business systems: first, the level of ownership integration, and second, the level of alliance integration (Whitley, 1999, 2007). Within these variables, he uses the following eight dimensions to systematically compare national business systems: (i) the means of owner control (direct, alliance, or market); (ii) the extent of production chain integration via ownership (low, medium, and high); (iii) the extent to which industrial sectors are integrated through ownership; (iv) the extent of alliance or horizontal coordination of production chains; (v) the extent to which competitors collaborate with one another; (vi) the extent of alliance coordination of sectors; (vii) the extent of interdependence among employers and employees; and (viii) the extent to which employees are trusted and delegated to within a firm (Whitley, 1999; Jackson & Deeg, 2006, 2008). This approach results in a typology of eight distinct forms of national business systems.[8] But beyond this, Whitley identifies and compares the core institutional features encouraging or restraining different varieties of business systems.[9]

One of the distinctive aspects of Whitley's approach is that it places organization and management at the centre of the analysis, as opposed to the structuring processes of broader economic and political phenomena (Morgan, 2007; for a more detailed overview of Whitley's perspective on "divergent capitalisms", see Whitley, 1999). From an institutional vantage point, Whitley aims to construct a clear methodological framework that, in addition to investigating training, collective bargaining, and workplace organization, recognizes the importance of the state, law, and regulation as they pertain to economic development, the institutions of ownership and finance, and the influence of family structure (Whitley, 1999; Morgan, 2007).

Whitley's (1999, 2007) typology of capitalisms is considered by some to be more precise and sophisticated in its interpretation and conceptualization of national economic varieties than the dichotomous LME-CME distinction developed in the original VoC literature (Morgan, 2007; Jackson & Deeg, 2006, 2008). Whitley is also credited for distinguishing more clearly between the dimensions of the broader institutional context and the internal elements of firm organization (Jackson & Deeg, 2006). Whitley's (2005, 2007) more recent work also aims to address the issue of within-system diversity by suggesting that the character of the state (e.g., liberal, promotional, developmental, business corporatist) strongly shapes the degree of institutional diversity or coherence across national business systems.

Despite the abovementioned points, the NBS approach is similar to the VoC approach in its focus on questions regarding the nature of capitalist variation, the

[8] Whitley's framework (1999, 2007, 13–17) defines eight ideal-types of business systems: (1) *fragmented* business systems, in which there is a very low degree of ownership and horizontal coordination among firms; (2) *project network* systems, in which ownership integration is minimal but firms are specialized and often cooperate or share resources for mutually beneficial results; (3) *industrial districts* display relatively low levels of ownership integration as well, but horizontal coordination among firms is high; (4) *financial conglomerate* systems are characterized by a high scope of ownership, while horizontal coordination is weak; (5) *integrated conglomerate* systems are also defined by low horizontal integration among firms, but both the scope and degree of ownership are high; (6) *compartmentalized* systems are exemplified by the U.S. case, wherein large integrated firms compete with one another in the context of arm's-length market arrangements; (7) *collaborative* systems suggest a corporatist-style organization of competing firms within a sector; and (8) *highly coordinated* systems coordinate economic activities through an alliance type of ownership, an example of which is the Japanese keiretsu.

[9] The main institutions included in Whitley's framework are: states, financial systems, skill development and control, and trust and authority. *States* are compared along the following dimensions: strength of their coordinating and developmental role, strength of market regulation, and extent to which they incorporate intermediary organizations. *Financial systems* are compared according to the traditional dichotomy between capital-market versus bank-based systems. The dimension of *skill development and control* is compared in terms of the strength of public education/training, union strength, the dominant organizing principle of unions, and the degree of bargaining centralization. The measure of trust and authority is compared in terms of the extent of trust within formal institutions, the type of business setting, and the relative of strength of different types of authority – paternalistic, communitarian, or contractarian (see Whitley, 1999).

economic advantages and disadvantages that correspond to different capitalist models, as well as processes of institutional convergence or divergence in the face of ongoing processes of economic globalization (Deeg & Jackson, 2007). Furthermore, Whitley's (1999) attention to the role of path dependence and institutional complementarities tends to privilege patterns of stability over change – a general preoccupation that is shared by the larger CC literature. Yet, as some scholars point out, Whitley's NBS approach, by incorporating a larger role for the state in the analysis, holds potential mechanisms for change based on the internal diversity/coherence of firms stemming from different patterns of state intervention, legitimacy, and trust (Whitley, 2005, 2007; Jackson & Deeg, 2008). Bringing trade fairs into the analysis of national business systems could introduce another important mechanism for change, as such events have considerable potential to intervene in some of the key areas used in the NBS approach to distinguish between and compare different national business systems. Trade fairs play a key role in facilitating the horizontal coordination of production chains, for example, and, by creating multiple opportunities for F2F exchanges and different types of meetings or interactions with competitors, these events can greatly influence the extent to which competitors ultimately collaborate with one another.

Informed by a rich literature on East Asian capitalisms, the NBS typologies offer a less Eurocentric interpretation of capitalist diversity than is typically found in the CC literature. This work also enables greater precision in categorizing capitalist varieties, such as "coordinated economies" (using Hall and Soskice's terminology), and more fine-grained distinctions among different forms of non-market coordination (Jackson & Deeg, 2006, 2008). Yet, as we have seen, there is a strong similarity between the NBS approach and the larger CC literature with respect to the overarching questions driving the research agenda and the basic assumptions underlying the approach. With these unifying elements in mind, we now turn to the final body of literature that is included in the broader CC tradition – namely, Economic Governance (EG) approaches.

3.3.4 *Economic Governance Approaches (EG)*

Economic Governance (EG)[10] approaches represent another, albeit less tightly knit, body of work that may be situated within the broader CC tradition. This literature includes studies based on the concept of social-systems-of-production (SSP) (e.g.,

[10]The use of the term "economic governance" to describe these approaches is adapted from Casey's (2009) work on the study of comparative capitalisms, which includes the development of an empirical CC index. This less unified strand of the CC literature has, however, been referred to in different ways by different scholars. In Jackson and Deeg's work (2006, 2008), for instance, the simple term "governance" is used to describe these approaches to comparative capitalism. Hall and Soskice (2001, 3) refer to this body of work as a "social systems of production approach", although they include other literatures under this rubric, such as analyses of national innovation systems and national business systems, thereby creating a much larger and more heterogeneous category of approaches to the study of comparative capitalism. While I agree that the NSI and NBS approaches share a number of common elements that define the CC literature as a whole, the present study distinguishes these approaches and discusses them separately from the "economic governance" or "governance" approaches.

Hollingsworth et al., 1994; Hollingsworth & Boyer, 1997), for example, as well as research following a flexible specialization style of analysis (Piore & Sabel, 1984). The origins and development of this more diverse body of work can be traced back to the French "regulation school", which is also concerned with variations in national institutions and the ways in which such differences affect patterns of growth, economic stability, and competitiveness at the level of the national state (Boyer, 1990). During the 1980s and 1990s, such approaches were further developed, most notably in the collaborative works of Hollingsworth et al. (1994), Crouch and Streeck (1997) and Hollingsworth and Boyer (1997). These collaborative works analyze the question of institutional diversity as it corresponds to six basic coordination mechanisms that are used in the governance of economic activity – markets, hierarchies, communities, the state, networks, and associations. These six governance mechanisms are said to vary along two basic dimensions: (i) the extent to which economic agents are guided by self-interest as opposed to a sense of social obligation; and (ii) the extent to which power follows a horizontal or vertical trajectory (Jackson & Deeg, 2008).

In keeping with the broader CC literature, the approach developed in these works suggests that each governance mechanism has a unique organizational structure and follows its own rules of exchange and administration (Hollingsworth & Boyer, 1997). Yet, unlike the VoC's focus on the rational, strategic action of economic agents within stable institutional settings, EG approaches generally stress the role of social norms and values in determining the behaviour of economic agents and their associated institutional outcomes (Jackson & Deeg, 2006).

While there is considerable diversity within the EG literature with respect to the level of analysis employed by scholars, the methodologies used to compare national economic systems, and what are deemed to be the distinguishing characteristics of capitalist varieties (for an extended discussion, see Jackson & Deeg, 2006, 2008), EG approaches nevertheless support the core assumptions that define the CC literature as a whole. The idea that economic action is inherently social due to the embeddedness of institutions in complex social settings is, for example, central to the EG scholarship (Hollingsworth & Boyer, 1997). EG approaches also emphasize the notions of institutional complementarities insofar as such research describes the interconnectedness of five main institutional types – namely, a wage-labour nexus, a configuration for competition, a monetary regime, a set of state interventions, and an international regime.

As mentioned above, the social systems of production (SSP) is a more recently refined perspective that falls under the rubric of EG approaches. The work of Hollingsworth and Boyer (1997), in particular, helped to develop this approach into a more complete and systematic type of analysis. A social system of production defines the way in which a variety of national institutions, organizations, and social norms or values are integrated into a particular social configuration. Such institutions include: the industrial relations system; educational and training system; the internal structure of firms; inter-firm relations; financial markets; the state and its policies; as well as a society's unique customs, traditions, norms, rules, and laws (Hollingsworth & Boyer, 1997, 2). The main thesis of the SSP perspective is that these different aspects are fundamentally intertwined and that to understand the performance of the entire system, one must appreciate the complex interconnections among its various components. In this way, it subscribes to the same basic tenet as the VoC approach and the broader CC tradition regarding the interdependency of institutions and the associated comparative advantages. Moreover, there is a similar adherence to the notion of path dependence in the underlying argument that the successful importation of new elements to the system will depend upon their interaction with other system elements (Hollingsworth & Streeck, 1994). The SSP approach highlights the central role of firms in economic action, but it links the organization of production to the supportive role played by economic institutions at different scalar levels of the political economy – e.g., subnational, transnational, global – arguing that "actors are increasingly nested in institutional arrangements which are linked at all levels" (Hollingsworth, 1998, 482). In this way, the SSP approach differs from the VoC approach, which adopts an exclusive focus on economic institutions operating at the level of the national state.

The "flexible specialization" approach, first developed by Piore and Sabel (1984) and later extended and reviewed by others (Schoenberger, 1987, 1988; Gertler, 1988), also fits under the broad EG rubric. Similar to the other CC approaches discussed above, this perspective views the emergence and long-term survival of production systems as dependent upon the extent to which such systems are compatible with the broader set of social relations in which they are embedded. This set can include industrial relations, skill formation, as well as certain forms of state intervention. At the core of this approach is the distinction between "mass production" and "craft production" or "flexible specialization" as technological paradigms (Hirst & Zeitlin, 1997).[11] Although the approach does not aim to advance one model over the other, it does suggest that the technological dynamism and developmental potential of each depends upon "a contingent and variable framework of institutional regulation at the microlevel of the firm or region and the macrolevel of the national and international economy" (Hirst & Zeitlin, 1997, 221). Each model follows its own logic with a view to producing sustainable productivity improvement and economic growth (Hollingsworth & Boyer, 1997). And, the institutional and environmental contexts are instrumental in determining the economic potential of the distinct models.

Also situated under the broad heading of EG approaches is Amable's (2003) more recent work, which aims to supplement the strategic interaction and equilibrium analysis of Hall and Soskice (2001) with a deeper appreciation of the type of social embeddedness and institutional choices conceptualized in the social systems of production framework (Jackson & Deeg, 2008). Based on five institutional spheres, Amable (2003) uses a cluster analysis to create a typology of five distinct models of capitalism.[12] Amable's (2003) work builds on the original VoC and SSP frameworks by further exploring the political aspects of capitalist systems, a

[11] In this context, the term *mass production* refers to the "manufacture of standardized products in high volumes using special-purpose machinery and predominantly unskilled labor" (Hirst & Zeitlin, 1997, 221). The term *craft production* or *flexible specialization*, on the other hand, is used to describe the "manufacture of a wide and changing array of customized products using flexible, general-purpose machinery and skilled, adaptable workers" (Hirst & Zeitlin, 1997, 221).

[12] Amable's (2003) typology is based on five institutional spheres of activity – (i) product market competition; (ii) labour market institutions; (iii) finance and corporate governance; (iv) social protection/welfare state; and (v) the education/training system. The five models of capitalist organization identified in his work are: (i) the market-based model; (ii) the social democratic model; (iii)

contribution that goes some way in addressing the general limitations of the CC literature regarding the evolution of capitalist models (for an extended discussion of this contribution, see Jackson & Deeg, 2006, 2008).

The subject of trade fairs has not been explored within the context of the EG literature; however, it is not difficult to see how these events might be viewed as a useful tool in supporting flexible systems of production through, for instance, the opportunities they provide for close interactions between firms and customers and their role in facilitating the acquisition of information about consumer trends and preferences. It would also be interesting to explore how trade fairs might influence different social systems of innovation and production (Amable, 1999; Amable et al., 1997), especially in areas of "products" and "competition", for example.

3.3.5 *Limitations of the CC Literature*

Despite its widespread influence in the field of comparative political economy, the CC literature, in general, and the VoC approach, in particular, has been the subject of considerable academic debate and criticism.[13] Without denying the significance

[13] For example, some critics take issue with the focus on the national state, arguing that different levels of analysis might reveal important elements of "within-system" diversity (Coates, 2005; Crouch, 2005; Panitch & Gindin, 2005). In the case of Germany, for example, Wojcik (2002) finds significant differences in the structure of corporate finance, not only between different sectors but also between different *Länder* (Peck & Theodore, 2007).

Other scholars criticize the VoC approach for failing to adequately deal with variation among different firms within national models (Allen, 2004; Crouch & Farrell, 2004). Some empirical work shows that economic practices may differ between firms in different as well as within the same sectors (Faulconbridge, 2008). In this context, I do not find it necessary to interpret the approach in such a way that it only allows for a single form of organization or a single set of practices. This criticism is part of a broader critique of macro-economic conceptions and their tendency to over-generalize.

Others suggest that the VoC framework is inclined towards institutional determinism in that it embraces an overly mechanistic conception of institutional complementarities, downplaying underlying power structures, such as social class and gender (Thelen, 2003; Crouch & Farrell, 2004; Coates, 2005; Pontusson, 2005; Jackson & Deeg, 2006). This is an important point; yet, I still believe that the approach makes a considerable contribution to the extent that it underscores the complementary nature of core institutions.

Empirical work using the VoC approach further suggests that the basic distinction drawn between LMEs and CMEs is overly simplistic (Howell, 2003; Streeck, 2005) and that successful applications of the framework tend to ignore cases that do not easily fit this typology (Haddow, 2008). The political economies of France, Italy, Spain, and Greece, for example, critically diverge from either ideal-type, leading to a chorus of calls for refinement of the initial typology or the construction of alternative models (Whitley, 1999; Amable, 2003; Lorenz & Lundvall, 2006; Schmidt, 2007). Statistical analyses have also shown that the classification of countries as LMEs or CMEs is not stable, and may change over time (Ahlquist & Breunig, 2009).

For other critics, the five core institutional arenas defined by the VoC approach are incomplete (Peck & Theodore, 2007). Indeed, when one considers issues such as the reproduction of the workforce and skill development, the social security system represents an additional institutional realm that warrants further conceptual development (Bathelt & Gertler, 2005). As Haddow (2008) of this broader body of criticism, this study concentrates on two main lines of criticism that are particularly pertinent to the present discussion. The first relates to the static nature of much of the CC literature and, consequently, the challenges it faces when it comes to explaining the more dynamic aspects of change and evolution in economic institutions (Crouch & Farrell, 2004; Crouch, 2005; Hancké & Goyer, 2005; Streeck & Thelen, 2005; Jackson & Deeg, 2006; Deeg & Jackson, 2007). Some recent studies have made important breakthroughs on this front by focusing on the incremental, yet cumulative and sometimes transformative ways in which institutional change occurs (Thelen, 2003; Thelen, 2004; Hall & Thelen, 2009; Mahoney & Thelen, 2010). The present study aims to build on this work by systematically exploring potential avenues for such dynamism across capitalist varieties, specifically in relation to inter-firm interaction at trade fairs.

The second line of criticism relates to the ambiguity surrounding the linkages between national states and the many potential forces of convergence and globalization (Crouch & Farrell, 2004; Pontusson, 2005; Panitch & Gindin, 2005; Peck & Theodore, 2007). While supporting Hall and Thelen's (2009, 9) assertion that the CC approach is not "incompatible with dynamic views of the political economy," it must also be recognized that the development of a more dynamic theory of capitalist varieties requires further clarification of the complex interplay between micro-, meso-, and macro-level phenomena (Deeg & Jackson, 2007). With this in mind, the present study aims to clarify the potential connections between micro-level firm search activities, on the one hand, and macro-level national patterns of technological specialization on the other. This is done by exploring the nature of cross-national inter-firm interactions at leading business-to-business trade fairs, with a view to better understanding how these interactions might influence the choices of firms with respect to following a pathway of convergence or divergence in their national technological specializations. With this in mind, I now turn to the longstanding problem of institutional change as it has been articulated and debated in the field of comparative politics.

3.4 The Problem of Institutional Change

Students of comparative politics – and, indeed, the social sciences more broadly – have long been interested in questions regarding the nature and dynamics of institutional change. Such questions have certainly not waned. If anything, they may have become even more salient in the context of the dramatic shifts and fluidity associated with a globalizing economy. Within this context, important questions are being raised about the capacity of existing institutions to creatively adapt and respond to new and emerging challenges and the manner in which they are doing so (Gertler & Wolfe, 2002; Pauly & Coleman, 2009).

Until quite recently, much of the literature on institutional change has focused on the role of exogenous shocks in bringing about radical change in institutions. Change initiated by endogenous developments that follow a more incremental process has, in contrast, been largely neglected (Mahoney & Thelen, 2010). The potential significance of these slower, more gradual shifts in institutions has, consequently, received relatively little attention in the broader institutionalist literature.[14] Yet, in the past several years, there has been a growing scholarly interest in, and more systematic attempts to investigate, the nature and consequences of gradual institutional change (e.g., Thelen, 2004, 2009; Streeck & Thelen, 2005; Mahoney, 2010; Mahoney & Thelen, 2010). Streeck and Thelen (2005), for example, draw from the theoretical literature on path dependence and its associated debates to develop a detailed inventory of frequently observed patterns of incremental institutional change. This work provides a useful framework for classifying empirical cases and then comparing them across different settings.

Yet, the development of such inventories or typologies is only a first step in constructing a more dynamic theory of institutional change. As Mahoney and Thelen (2010, 3) point out, "institutional analysts must go beyond classification to develop causal propositions that locate the sources of institutional change – sources that are not simply exogenous shocks or environmental shifts". In their recent edited volume (Mahoney & Thelen, 2010), Mahoney and Thelen go much further to create a model of institutional change that connects specific processes of incremental change to the character of the institutional setting, as well as the properties of the institutions themselves that enable particular types of "change strategies" and "change agents". This is an essential step in the development of a more complete theory of institutional change and one that goes a long way to address some of the open questions underlying the sources and varieties of endogenous institutional change.[15] But there are still many issues that remain unresolved.[16] We still do not know, for instance, how and why processes of institutional change vary across issue areas or domains of the political economy. Whereas some areas may generally follow a slow, gradual process of institutional transformation, change in other domains may typically be more abrupt and wholesale. Useful insights could also be gained through systematically

[14] As Mahoney (2010) observes, it is problematic to treat incremental change simply as the opposite of punctuated change: "if punctuated change is both a big change and a fast change, it does not automatically follow that an incremental change is both a little change and a slow change" (Mahoney, 2010, 8). Indeed, the concept of incremental change is also used to describe "*a series of small changes that add up to a big change over a long period of time*" (Mahoney, 2010, 8 [emphasis in original]). As Mahoney's (2010) work demonstrates, incremental change, of both the continuous and non-continuous variety, can match punctuated change in terms of its ultimate significance.

[15] Such questions include the following (see Mahoney & Thelen, 2010, 3): What are the precise institutional properties that allow change? How and why do such properties permit or enable economic agents to carry out change activities or behaviours? What, exactly, are such "change activities" or behaviours? How might we conceptualize these "change agents? Do different types of strategies thrive/languish in different institutional settings? And what aspects of the institutions themselves render them more or less amenable to particular change strategies?

[16] These open questions are based on Hall's (2010, 219–220) reflections on the problem of institutional change and recent efforts, particularly within the historical institutionalist tradition, to develop a more complete model of institutional change that includes both exogenous and endogenous stimuli.

comparing empirical cases of institutional change with empirical examples of institutional stability. In relation to institutional dynamism, Hall (2010) points out that there are many unanswered questions when it comes to the nature and sources of innovation. On this latter point, I argue that international trade fairs offer a particularly useful vantage point from which to explore such questions. Moreover, I suggest that these events can be viewed as a potentially important instrument through which institutional change might be channeled or even act as a trigger for institutional change.

It is within the context of the above developments that the present research aims to garner a better understanding of the character of endogenously supplied "change agents", as well as the particular types of "change strategies" that may flourish or dominate in a specific institutional setting – namely, international trade fairs. I begin by reviewing the challenges faced by the three leading approaches to institutional analysis – i.e., rational-choice, sociological, and historical institutionalism – with regards to explaining the dynamic elements of institutional change (Mahoney & Thelen, 2010). Given that detailed descriptions of the different perspectives exist elsewhere (see, for example, Hall & Taylor, 1996), this chapter focuses specifically on how each approach typically addresses the question of institutional change and the particular difficulties or advantages it has in doing so. In general, this review finds that, despite their conceptual and methodological differences, all three approaches are similar in the sense that they are better equipped to explain why and how institutions endure over time, as opposed to identifying the mechanisms by which endogenously-driven change occurs. It is only quite recently that a small group of scholars has attempted to deal with this shortcoming in the literature in a systematic and comprehensive way (see, in particular, Thelen, 2003, 2004; Streeck & Thelen, 2005; Mahoney & Thelen, 2010).

The present study uses the literature on institutional change as a way of analyzing the potential role of international trade fairs in the globalizing knowledge-based political economy. None of the perspectives discussed below explores the role of international trade fairs. Yet, this chapter looks at how the different approaches might incorporate these events and how trade fairs, defined as "temporary clusters" or "temporary markets", could offer a useful window to the nature and dynamics of a specific type of "change agent" (Mahoney & Thelen, 2010) – namely, firms from different capitalist systems. Further, it suggests that trade fairs could be conceptualized as agents of change in their own right because their defining characteristics provide participants with "critical openings for creativity and agency" (Mahoney & Thelen, 2010, 12), which may, in turn, lead to gradual changes in the technological practices and processes that ultimately shape institutions.

Although there is great diversity in the ways in which institutions are treated in the relevant academic literature, from a definitional standpoint, at least, there is widespread agreement that institutions constitute relatively enduring components of political and social life. This notion of persistence, which is tied to most conceptions of institutions, has not surprisingly led to a greater emphasis on explaining patterns of continuity than change (Mahoney & Thelen, 2010). This is seen across all three institutionalist approaches. That being said, there are some important differences between the various approaches that influence their explanatory capacity with respect to the question of institutional change. This is particularly true with respect to the more dynamic elements of change in institutional settings. With this objective in mind, a brief description of the different approaches is offered below.

3.4.1 Rational-Choice Institutionalism

From a rational-choice perspective, institutions are understood as "patterns of regularized behavior that reflect Pareto-optimal equilibria or subgame perfect solutions to collective actions dilemmas" (Hall, 2010, 204). In other words, rational-choice institutionalists view institutions as coordinating mechanisms that help sustain certain equilibria (Shepsle, 1989; Calvert, 1995; Levi, 1997; Mahoney & Thelen, 2010). Such institutionalized patterns of behaviour are said to remain steady "because the actors cannot improve their positions by defecting from the pattern of behavior" (Hall, 2010, 204; see also, Shepsle, 1989; Greif & Laitin, 2004). Whereas rational-choice institutionalists have provided important insights into the workings of institutions and the processes leading to their persistence over time, explanations of institutional change within this tradition have largely been based on a model of discontinuous change, wherein institutional change only takes place when an exogenous shock alters or disrupts the institutional setting (Hall, 2010). By focusing on "comparative statics" (Levi, 2008), the rational-choice perspective on institutional change has great difficulty accounting for incremental processes of endogenous change in institutions. And, while there have been some attempts to address this problem within the rational-choice tradition (see, in particular, Greif & Laitin, 2004; see, also, Thelen's commentary on their approach 2004, 30), the bulk of the literature continues to sidestep the issue of endogenously generated, incremental forms of institutional evolution, rather than seek new ways to inject greater dynamism into the analysis of institutional change.

Although trade fairs have not been discussed in this literature, these events could represent an important new way of introducing greater dynamism into the rational-choice institutionalist's account of institutional change. In terms of methodological advantages alone, proponents of all three institutionalist perspectives will find that trade fairs are a particularly valuable area of inquiry because they provide an opportunity for researchers to observe, first hand, in "tractable settings bounded by time and space", the "sense-making and sense giving processes" underlying processes of dynamic change. (Lampel & Meyer, 2008, 1030). But, specific to rational-choice institutionalists, trade fairs expose institutional actors – in a temporary but intense way – to new sets of institutional and technological practices or conditions that can "overwhelm the self-enforcing behavior" (Greif & Laitin, 2004) of the existing institutions in which they operate.

3.4.2 *Sociological Institutionalism*

Originating within the subfield of organization studies, sociological institutionalism adopts a broader conceptualization of institutions than either rational-choice or historical institutionalists. Defining institutions as "the symbol systems, cognitive scripts, and moral templates that provide the 'frames of meaning' guiding human action", sociological institutionalists essentially blur the conceptual distinction between "culture" and "institutions" (Hall & Taylor, 1996, 947). By treating culture as an institution, sociological explanations of institutional change necessarily also involve changes to these wider structures of meaning, cognitive scripts, and so on. In addressing the problem of institutional change, proponents of this approach frequently draw attention to the self-reproducing properties of institutions (Powell, 1991; Jepperson, 1991; Mahoney & Thelen, 2010). Some scholars within this tradition connect the self-reproducing properties of institutions to "codes of appropriateness" that become deeply ingrained in institutional agents through processes of learning and socialization (March & Olsen, 1984; Mahoney & Thelen, 2010). Others see such properties as more cognitive in nature, where the institutions are embedded in the social context to such a degree that they are "taken for granted" and self-reproduction occurs outside the realm of conscious scrutiny (Zucker, 1983, 2; Mahoney & Thelen, 2010). Institutional reproduction can also take place through the process of "institutional isomorphism", which has been described as "a constraining process that forces one unit in a population to resemble other units that face the same set of environmental conditions" (DiMaggio & Powell, 1983, 149; Hawley, 1968). In this process, agents tend to reproduce the same or compatible organizational logics across different spheres by preserving and carrying forward existing frames of meaning when constructing new institutions – even if such logics have become outdated or lost their utility (Mahoney & Thelen, 2010, 5).

Sociological institutionalists view the relationship between institutions and individual action quite differently from other institutionalist perspectives. They see this relationship as highly interdependent and mutually-constitutive (Hall & Taylor, 1996), with institutions establishing the boundaries of individual action by shaping the wider frameworks of meaning and belief systems within which they operate. In this respect, institutions may be said to "create the lenses through which actors see the world and the very categories of structure, action, and thought" (DiMaggio & Powell, 1991, 13; Hall & Taylor, 1996). Such categories are critical to individual action, as they provide agents with the interpretive frameworks necessary to make sense of the world around them and the behaviour of others (Hall & Taylor, 1996). In this way, "institutions do not simply affect the strategic calculations of individuals, as rational choice institutionalists contend, but also their most basic preferences and very identity" (Hall & Taylor, 1996, 948).

Within this broad research tradition, sociological institutionalists have identified a powerful set of mechanisms to explain patterns of institutional reproduction and stability. Yet, these mechanisms offer relatively little insight into the more incremental and dynamic processes of endogenous institutional change (Mahoney &

Thelen, 2010). Thus, similar to rational-choice institutionalists, sociological institutionalists have a hard time accounting for institutional change without falling back on some type of exogenously-derived factor – such as, the introduction of new interpretive frames from an external source (DiMaggio & Powell, 1983; Hannan & Freeman, 1989; Mahoney & Thelen, 2010). Trade fairs seem relevant on this front because they are endogenous to the institution in the sense that they represent an internal choice or a foundational part of the organization's operations – for example, as part of its marketing mix, its R&D activities, its innovation strategy, or its market outreach activities. They are also a platform for information and knowledge exchange that is characterized by a shared industry-based language and a common industry culture or value system. Yet, at the same time, trade fairs are a critical mechanism by which the intensive interaction of global actors at these events brings about opportunities to reshape long-held interpretive frameworks through, for instance, "the expansion of cognitive horizons" – to employ Ruggie's expression (1993, 155).

3.4.3 *Historical Institutionalism*

Compared to its sociological counterpart, historical institutionalism generally adopts a narrower view of institutions as "the formal or informal procedures, routines, norms and conventions embedded in the organizational structure of the polity or political economy" (Hall & Taylor, 1996, 938). While recognizing their potential coordinating functions as well as their cultural aspects, historical institutionalists see institutions primarily as "the political legacies of concrete historical struggles" (Mahoney & Thelen, 2010, 7; for an extended discussion, see Thelen, 1999, 381–99). From this it seems to follow that path dependent explanations, which include the central concepts of "legacy" and "critical junctures" (Mahoney, 2001), have traditionally assumed an important place in historical institutionalist accounts of institutional change or, more often, continuity.[17] The historical institutionalist interpretation of institutions also suggests that most scholars within this tradition adopt a power-distributional approach to institutional analyses, with many accounts of institutional continuity based on the dynamic of increasing returns to power (Mahoney & Thelen, 2010).

How, exactly, do historical institutionalists deal with the problem of institutional change? As indicated above, until recently, they too have been more adept at explaining processes of institutional continuity than change (Mahoney & Thelen, 2010).

[17] In comparative politics, the notion of path dependence suggests that "once a country or region has started down a track, the costs of reversal are very high. There will be other choice points, but the entrenchments of certain institutional arrangements obstruct an easy reversal of the initial choice" (Levi, 1997, 28). Thus, even if the institution's original rationale becomes irrelevant over time due to other external changes, the legacy of past choices may often preclude a corresponding institutional change because of the costs associated with readjustment.

Traditional interpretations have largely relied on the concept of "critical junctures" as a way of accounting for change in institutions. Such arguments are typically based on a dual model of institutional evolution involving relatively long stretches of path dependent institutional stability that are periodically disrupted by short, sudden episodes of institutional flux – or "critical junctures" (Capoccia & Keleman, 2007). Despite their brevity, these critical junctures denote periods of time that allow more dramatic change to take place (Katznelson, 2003). According to this line of reasoning, the decisions made during critical junctures can have a profound and enduring impact on the development of an institution because they shut off other possible development paths and encourage institutions that create self-reproducing, path dependent processes (Capoccia & Keleman, 2007). As Pierson (2004, 135) explains, "[j]unctures are 'critical' because they place institutional arrangements on paths or trajectories, which are very difficult to alter." In this respect, the historical institutionalist approach resembles other punctuated equilibrium models in that it relies on a discontinuous model of institutional change. By conceptualizing change in terms of institutional collapse and replacement, such models provide an incomplete picture of institutional change – one that disregards more gradual, endogenous processes of change in institutions (Mahoney & Thelen, 2010).

In recent years, however, there has been a movement within the historical institutionalist literature, in particular, that aims to shift scholarly attention away from the standard punctuated equilibrium model and towards more dynamic and incremental processes of endogenous institutional change (Thelen, 2003; Streeck & Thelen, 2005; Mahoney, 2010; Mahoney & Thelen, 2010). This movement represents a critical step in the development of a more comprehensive model of institutional change. In this way, it may be seen as an initial or, at least, partial response to recent calls for a more dynamic approach to the study of capitalist diversity (Deeg & Jackson, 2007; Jackson & Deeg, 2008).

Both individually (Thelen, 2003, 2004, 2009) and in collaboration with other prominent scholars, such as Wolfgang Streeck and James Mahoney (see, for example, Streeck & Thelen, 2005; Mahoney & Thelen, 2010), Thelen's work stands out as a clear challenge to the dominant institutionalist explanations of change. This is done through the introduction and development of new conceptual tools to systematically explore and analyze incremental, endogenously-derived processes of institutional change. Without denying the continued utility of existing conceptual devices – namely, "critical junctures" and "path dependence" – to account for abrupt patterns of institutional change followed by long periods of stability and reproduction, Thelen (2003, 2004, 2009) persuasively argues that such tools do not adequately capture all, or even the most typical ways that institutions change over time. On the contrary, she demonstrates that, in many cases, "significant change often takes place gradually and through a cumulation of seemingly small adjustments" (Thelen, 2009, 475). Thus, at the very least, the prevailing punctuated equilibrium model of institutional change (i.e., breakdown/replacement) needs to be supplemented by a more comprehensive analytical framework that can capture and also guide us in analysing processes of institutional change that are incremental but cumulatively transformative in nature (Thelen, 2009; Mahoney & Thelen, 2010).

Together, the work of Streeck and Thelen (2005) and Mahoney and Thelen (2010) provide the core components of a model that aims to capture the more dynamic and incremental processes of endogenous institutional change. The basic framework for explaining different modes of incremental change (i.e., displacement, layering, drift, and conversion)[18] involves, first, the characteristics of both the political context and the institution, which together are said to determine the particular form of institutional change (Mahoney & Thelen, 2010, 15). The political and institutional contexts are able to achieve this through their role in shaping the type of change agent that is expected to arise and gain traction in a given context. The dynamic relationship between the characteristics of the political and institutional context, the type of change agent, and the nature of institutional change are further linked through the elaboration of distinct causal variables – namely, strong/weak veto possibilities in the political arena, low/high levels of discretion in the targeted institution, and the specific character of the change agent (i.e., insurrectionaries, symbionts, subversives, opportunists).[19]

The present study does not aim to provide a detailed description or direct application of the model proposed by Mahoney and Thelen (2010); yet it supports the argument that exogenous shocks are not responsible for all, or even most, forms of institutional change. As such, it follows Mahoney and Thelen's (2010) example in searching for new ways to define the properties of institutions that reveal the dynamic element or elements underlying such change. Whereas the structure of this research project does not focus on a meaningful account of the actual modes of change produced through inter-firm interaction at trade fairs, it does enable a thorough investigation into the nature of a particular type of change agent (i.e., exhibiting firms) operating within these dynamic institutional settings and the types of strategies or behaviours they employ. It also suggests that international trade fairs themselves might be considered an important type of change agent in that they represent a key mechanism by which individual actors arrive at decisions, which, in turn, may influence the broader institutional environments in which they operate.

Although trade fairs have not been explored in this literature, they are a phenomenon worth considering through this lens. A prominent feature of historical institutionalism, for example, is its emphasis on power and asymmetrical power relations

[18] The different modes of institutional change are defined by identifying the particular locus of such change (Streeck & Thelen, 2005; Mahoney & Thelen, 2010). *Displacement*, for instance, simply involves the elimination of existing rules and the introduction of new ones; *Layering* describes the introduction of new rules in addition to existing ones; *Drift* implies a shift in the actual impact of existing rules as a result of broader environmental changes; and *Conversion* denotes a change in how existing rules are enacted as a result of "strategic redeployment".

[19] In Mahoney and Thelen's (2010, 22–27) work, the different types of change agents are classified based on how they correspond to the following two questions: (i) Does the actor seek to preserve the existing institutional rules? (ii) Does the actor abide by the institutional rules? Insurrectionaries seek neither to preserve nor follow the rules of an institution; symbionts seek to preserve but not follow institutional rules; subversives do not preserve but aim to follow institutional rules; and opportunists have ambiguous preferences, "exploiting whatever possibilities exist within the prevailing system to achieve their ends" (Mahoney & Thelen, 2010, 26).

(Hall & Taylor, 1996). Whereas power relations are an important aspect of institutional research generally, "historical institutionalists have been particularly attentive to the way in which institutions distribute power unevenly across social groups" (Hall & Taylor, 1996, 9). On this front, it could be argued that international trade fairs play something of a moderating role among actors who enjoy differing degrees of influence on the global economic stage. While the relative power of some firms (and their corresponding countries) compared to others is openly laid out at trade fairs (e.g., the size of the exhibit booths), these events bring together both well-known and less well-known firms from a particular industry or sector. What this means is that all trade fair participants have a fairly equal opportunity to experience the specific communication and information ecology – or "global buzz" (Maskell et al., 2006) – that is created during these events. This is not to say that differences in power are entirely leveled at trade fairs, or that asymmetries in power in the global marketplace, for instance, cannot be further reinforced through these events. Rather, it is to suggest that trade fairs are well-positioned to mediate existing relationships among firms from different political-economic contexts in ways that few other entities possess and in ways that may, ultimately, have profound implications for the broader institutions which govern these firms.

3.5 Conclusion

This chapter outlines the macro-level theoretical foundations of the study's forthcoming analysis of inter-firm interaction at trade fairs. Drawing heavily on the VoC, as well as the broader CC literature, it suggests that the specific arrangement of institutions operating within different capitalist varieties influences the behaviour of the firms that operate within them. In keeping with the CC literature, I suggest that these institutional configurations influence the search processes of firms at trade fairs so that, ultimately, the adoption or acquisition of new technological ideas, solutions, or business partners will more or less closely fit with the established production models found in the firms' home countries. In this way, the study supports the core assumptions of the VoC and broader CC literature. Yet, as will be described more fully in Chap. 4, when the focus shifts from the macro- to the micro-level, the technological search processes of firms rarely follow a rational or linear trajectory, despite being guided by their home context. Instead, the nature of firms' search and information processes at trade fairs more closely resembles the garbage-can processes operating within so-called "organized anarchies" (Cohen et al., 1972; Kingdon, 2003).

This study uses the literature on institutional change as a way of analyzing the role of international trade fairs in shaping the practices and processes associated with the technological specialization of firms within the context of the global knowledge economy. It suggests that scholars interested in developing more dynamic theories of institutional change in relation to capitalist varieties and the comparative study of such varieties would do well to turn their analytical lens to these events.

One reason for this is that international trade fairs, occupying the meso-level as "microcosms" of an emerging industry, marketplace, or technology, can help clarify the linkages between national states and the various processes that may be associated with convergence and economic globalization.

Finally, this chapter has argued that recent work on the subject of incremental, endogenously-driven processes of institutional change (e.g., Thelen, 2003, 2004, 2009; Streeck & Thelen, 2005; Mahoney, 2010; Mahoney & Thelen, 2010) represents a critical step in the development of a more dynamic theory of capitalist diversity. While there remain many open questions that warrant further attention in future work, the abovementioned contributions provide new conceptual tools that will aid in the refinement and systematic analysis of different modes and mechanisms of institutional change. Although the present study lacks the longitudinal dimension required to arrive at meaningful conclusions regarding the modes of change associated with inter-firm interaction at international trade fairs, it nevertheless builds on recent institutionalist developments by offering an in-depth empirical investigation into the nature of change agents and change strategies within a unique and, in this context, still largely under-examined institutional setting – international trade fairs. Before turning to this challenge in the empirical portion of the study, the next chapter concludes the theoretical discussion, focusing specifically on micro-level dynamics with the help of the garbage-can model of organizational choice.

References

Ahlquist, J. S., & Breunig, C. (2009). *Country clustering in comparative political economy*. Discussion Paper 09/5. Max-Planck-Institut Für Gesellschaftsforschung (Max Planck Institute for the Study of Societies).

Akkermans, D., Castaldi, C., & Los, B. (2009). Do "liberal market economies" really innovate more radically than coordinated market economies? Hall and Soskice reconsidered. *Research Policy, 38*, 181–191.

Allen, M. M. C. (2004). The varieties of capitalism paradigm: Not enough variety? *Socio-Economic Review, 2*, 87–108.

Amable, B. (1999). *Institutional complementarity and diversity of social systems of innovation and production*. WZB Discussion Paper, No. FS I: 99–309.

Amable, B. (2003). *The diversity of modern capitalism*. Oxford University Press.

Amable, B., Barré, R., & Boyer, R. (1997). Les systèmes d'innovation à l'ère de la globalisation. *Economica*.

Aoki, M. (2001). *Comparative institutional analysis*. MIT Press.

Archibugi, D., & Coco, A. (2004). A new indicator of technological capabilities for developed and developing countries (ArCo). *World Development, 32*(4), 629–654.

Archibugi, D., & Pianta, M. (1992). *The technological specialization of advanced countries: A report to the EEC on international science and technology activities*. Kluwer Academic Publishers.

Archibugi, D., Howells, J., & Michie, J. (1999). *Innovation policy in a global economy*. Cambridge University Press.

Bathelt, H., & Gertler, M. S. (2005). The German variety of capitalism: Forces and dynamics of evolutionary change. *Economic Geography, 81*, 1–9.

Bathelt, H., & Schuldt, N. (2008). Between luminaires and meat grinders: International trade fairs as temporary clusters. *Regional Studies, 42*(6), 853–868.

Bathelt, H., Golfetto, F., & Rinallo, D. (2014). *Trade shows in the globalizing knowledge economy*. Oxford University Press.

Biggart, N. W. (1991). Explaining Asian economic organization. Toward a Weberian institutional perspective. *Theory and Society, 20*, 199–232.

Boyer, R. (1990). *The regulation school: A critical introduction*. Columbia University Press.

Calvert, R. L. (1995). Rational actors, equilibrium, and social institutions. In J. Knight & I. Sened (Eds.), *Explaining social institutions* (pp. 57–93). University of Michigan Press.

Capoccia, G., & Keleman, D. R. (2007). The study of critical junctures: Theory, narrative, and counterfactuals in historical institutionalism. *World Politics, 59*(3), 341–369.

Casey, T. (2009). Mapping stability and change in advanced capitalisms. *Comparative European Politics, 7*, 255–278.

Chandler, A. D. (1978). *The visible hand: The managerial revolution in American business*. Harvard University Press.

Coates, D. (2005). Paradigms of explanation. In D. Coates (Ed.), *Varieties of capitalism, varieties of approaches* (pp. 1–25). Palgrave.

Cohen, M. D., March, J. G., & Olsen, J. P. (1972). A garbage can model of organizational choice. *Administrative Science Quarterly, 17*(1), 1–25.

Crouch, C. (2005). *Capitalist diversity and change: Recombinant governance and institutional entrepreneurs*. Oxford University Press.

Crouch, C., & Farrell, H. (2004). Breaking the path of institutional development? Alternatives to the new determinism. *Rationality and Society, 16*(1), 5–43.

Crouch, C., & Streeck, W. (1997). *Political economy of modern capitalism: Mapping convergence and diversity*. Sage.

Deeg, R., & Jackson, G. (2007). Towards a more dynamic theory of capitalist variety. *Socio-Economic Review, 5*, 149–179.

DiMaggio, P. J., & Powell, W. W. (1983). The iron cage revisited: Collective rationality and institutional isomorphism in organizational fields. *American Sociological Review, 48*(2), 147–160.

DiMaggio, P. J., & Powell, W. W. (1991). Introduction. In P. J. DiMaggio & W. W. Powell (Eds.), *The new institutionalism in organizational analysis* (pp. 1–38). University of Chicago Press.

Djankov, S., Glaeser, E. L., La Porta, R., Lopez-de-Silane, F., & Shleifer, A. (2003). The new comparative economics. *Journal of Comparative Economics, 31*(4), 595–619.

Dosi, G. (1988). The nature of the innovative process. In G. Dosi, C. Freeman, R. R. Nelson, G. Silverberg, & L. Soete (Eds.), *Technical change and economic theory* (pp. 210–238). Pinter.

Edquist, C. (Ed.). (1997). *Systems of innovation: Technologies, institutions and organizations*. Routledge.

Esping-Andersen, G. (1996). After the golden age? Welfare state dilemmas in a global economy. In G. Esping-Andersen (Ed.), *Welfare states in transition: National adaptations in global economies* (pp. 1–31). Sage.

Evangelista, R. (1999). *Knowledge and investment*. Edward Elgar Publishing.

Faulconbridge, J. R. (2008). Managing the transnational law firm: A relational analysis of professional systems, embedded actors, and time-space sensitive governance. *Economic Geography, 84*(2), 185–210.

Freeman, C. (1982). *The economics of industrial innovation* (2nd ed.). Frances Pinter.

Freeman, C. (1987). *Technology, policy and economic performance: Lessons from Japan*. Frances Pinter.

Freeman, C. (1995). The 'National system of innovation' in historical perspective. *Cambridge Journal of Economics, 19*, 5–24.

Gertler, M. S. (1988). The limits of flexibility: Comments on the post-Fordist vision of production and its geography. *Transactions of the Institute of British Geographers, 13*(4), 419–432.

Gertler, M. S. (2004). *Manufacturing culture: The institutional geography of industrial practice*. Oxford University Press.

Gertler, M. S., & Wolfe, D. A. (Eds.). (2002). *Innovation and social learning: Institutional adaptation in an era of technological change*. Palgrave Macmillan.

Granovetter, M. (1985). Economic action and social structure: The problem of embeddedness. *American Journal of Sociology, 91*, 481.

Greif, A., & Laitin, D. D. (2004). A theory of endogenous institutional change. *American Political Science Review, 98*(4), 633–652.

Haake, S. (2002). National business systems and industry-specific competitiveness. *Organization Studies, 23*(5), 711–736.

Haddow, R. (2008). How can comparative political economy explain variable change? Lessons for, and from, Canada. In L. A. White, R. Simeon, R. Vipond, & J. Wallner (Eds.), *The comparative turn in Canadian political science* (pp. 221–237). UBC Press.

Hall, P. A. (2010). Historical institutionalism in rational and sociological perspective. In J. Mahoney & K. Thelen (Eds.), *Explaining institutional change: Ambiguity, agency, and power* (pp. 204–224). Cambridge University Press.

Hall, P. A., & Soskice, D. (2001). An introduction to varieties of capitalism. In P. A. Hall & D. Soskice (Eds.), *Varieties of capitalism: The institutional foundations of comparative advantage* (pp. 1–68). Oxford University Press.

Hall, P. A., & Taylor, R. C. (1996). Political science and the three new institutionalisms. *Political Studies, 44*, 936–957.

Hall, P. A., & Thelen, K. (2009). Institutional change in varieties of capitalism. *Socio-Economic Review, 7*(1), 7–34.

Hancké, B., & Goyer, M. (2005). Degrees of freedom: Rethinking the institutional analysis of economic change. In G. Morgan, R. Whitley, & E. Moen (Eds.), *Changing capitalisms? Internationalism, institutional change, and systems of economic organization* (pp. 53–77). Oxford University Press.

Hancké, B., Rhodes, M., & Thatcher, M. (Eds.). (2007). *Beyond varieties of capitalism: Conflict, contradiction, and complementarities in the European economy*. Oxford University Press.

Hannan, M. T., & Freeman, J. (1989). *Organizational ecology*. Harvard University Press.

Hawley, A. (1968). Human ecology. In D. L. Sills (Ed.), *International encyclopedia of the social sciences* (pp. 328–337). Macmillan.

Hirst, P., & Zeitlin, J. (1997). Flexible specialization: Theory and evidence in the analysis of industrial change. In J. R. Hollingsworth & R. Boyer (Eds.), *Contemporary capitalism: The embeddedness of institutions* (pp. 220–239). Cambridge University Press.

Hollingsworth, J. R. (1998). On multi-level analysis. In J. R. Hollingsworth, K. H. Müller, & E. J. Hollingsworth (Eds.), *Advancing socio-economics: An institutionalist perspective* (pp. 19–36). Rowman & Littlefield Publishers, Inc.

Hollingsworth, J. R., & Boyer, R. (Eds.). (1997). *Contemporary capitalism: The embeddedness of institutions*. Cambridge University Press.

Hollingsworth, J. R., & Streeck, W. (1994). Countries and sectors: Performance, convergence, and competitiveness. In J. R. Hollingsworth, P. C. Schmitter, & W. Streeck (Eds.), *Governing capitalist economies: Performance and control of economic sectors* (pp. 270–230). Oxford University Press.

Hollingsworth, J. R., Schmitter, P. C., & Streeck, W. (1994). *Governing capitalist economies: Performance and control of economic sectors*. Oxford University Press.

Howell, C. (2003). Varieties of capitalism: And then there was one? *Comparative Politics, 34*, 103–124.

Jackson, G., & Deeg, R. (2006). *How many varieties of capitalism? Comparing the comparative institutional analyses of capitalist diversity*. MPIfG Discussion Paper 06/2, Max Planck Institute for the Study of Societies.

Jackson, G., & Deeg, R. (2008). Comparing capitalisms: Understanding institutional diversity and its implications for international business. *Journal of International Business Studies, 39*(4), 540–561.

Jepperson, R. L. (1991). Institutions, institutional effects, and institutionalism. In W. W. Powell & P. J. DiMaggio (Eds.), *The new institutionalism in organizational analysis* (pp. 143–163). University of Chicago Press.

Johnson, C. (1982). *MITI and the Japanese miracle: The growth of industrial policy: 1925–1975*. Stanford University Press.

Katznelson, I. (2003). Periodization and preferences: Reflections on purposive action in comparative historical science. In J. Mahoney & D. Rueschmeyer (Eds.), *Comparative political historical analysis: Achievements and agendas* (pp. 270–304). Cambridge University Press.

Kingdon, J. W. (2003). *Agendas, alternatives and public policies* (2nd ed.). Addison-Wesley Educational Publishers.

Lampel, J., & Meyer, A. D. (2008). Guest editors' introduction: Field-configuring events as structuring mechanisms: How conferences, ceremonies, and trade shows constitute new technologies, industries, and markets. *Journal of Management Studies, 45*(6), 1025–1035.

Levi, M. (1997). A model, a method, and a map: Rational choice in comparative and historical analyses. In M. Irving Lichbach & A. S. Zuckerman (Eds.), *Comparative politics: Rationality, culture, and structure* (pp. 19–41). Cambridge University Press.

Levi, M. (2008). Reconsiderations of rational choice in comparative and historical analysis. In M. I. Lichbach & A. Zuckerman (Eds.), *Comparative politics: Rationality, culture, and structure* (2nd ed., pp. 117–133). Cambridge University Press.

Lipietz, A. (1987). *Mirages and miracles: Crisis in global Fordism*. Verso.

Lorenz, E., & Lundvall, B.-Å. (Eds.). (2006). *How Europe's economies learn: Coordinating competing models*. Oxford University Press.

Lundvall, B.-Å. (1985). *Product innovation and user-producer interaction*. Aalborg University Press.

Lundvall, B.-Å. (1992). *National systems of innovation*. Pinter.

Lundvall, B.-Å., & Freeman, C. (1988). *Small countries facing the technological revolution*. Frances Pinter Publishers Ltd..

Lundvall, B.-Å., & Maskell, P. (2000). Nation states and economic development: From national systems of knowledge creation and learning. In G. L. Clark, M. P. Feldman, & M. S. Gertler (Eds.), *The Oxford handbook of economic geography* (pp. 353–372). Oxford University Press.

Mahoney, J. (2001). Path-dependent explanations of regime change: Central America in comparative perspective. *Studies in Comparative International Development, 36*(1), 111–141.

Mahoney, J. (2010). Conceptualizing and explaining punctuated versus incremental change. In *Paper prepared for the Annual Meetings of the American Political Science Association*.

Mahoney, J., & Thelen, K. (2010). A theory of gradual institutional change. In J. Mahoney & K. Thelen (Eds.), *Explaining institutional change: Ambiguity, agency, and power* (pp. 1–37). Cambridge University Press.

March, J. G., & Olsen, J. P. (1984). The new institutionalism: Organizational factors in political life. *The American Political Science Review, 78*(3), 734–749.

Maskell, P., & Malmberg, A. (1999). The competitiveness of firms and regions: 'Ubiquitification' and the importance of localised learning. *European Urban and Regional Studies, 6*(1), 9–25.

Maskell, P., Bathelt, H., & Malmberg, A. (2006). Building global knowledge pipelines: The role of temporary clusters. *European Planning Studies, 14*, 997–1013.

Metcalfe, S. (1997). Technology systems and technology policy in an evolutionary framework. In D. Archibugi & J. Michie (Eds.), *Technology, globalization and economic performance* (pp. 268–296). Cambridge University Press.

Morgan, K. (2007). The learning region: Institutions, innovation and regional renewal. *Regional Studies, 41*(1), S147–S159. https://doi.org/10.1080/00343400701232322

Nelson, R. R. (1988). Institutions supporting technical change in the United States. In G. Dosi, C. Freeman, R. R. Nelson, G. Silverberg, & L. Soete (Eds.), *Technical change and economic theory* (pp. 312–329). Pinter.

Nelson, R. R. (1993). *National innovation systems*. Oxford University Press.

Panitch, L., & Gindin, S. (2005). Euro-capitalism and American empire. In D. Coates (Ed.), *Varieties of capitalism, varieties of approaches* (pp. 139–159). Palgrave Macmillan.

Patel, P. (1997). Localized production of technology for global markets. In D. Archibugi & J. Michie (Eds.), *Technology, globalization and economic performance* (pp. 198–214). Cambridge University Press.

Patel, P., & Pavitt, K. (1991). Large firms in the production of the world's technology: An important case of "non-globalisation". *Journal of International Business Studies, 22*(1), 1–21.

Pauly, L. W., & Coleman, W. D. (Eds.). (2009). *Global ordering: Institutions and autonomy in a changing world*. UBC Press.

Pavitt, K. (1998). Technologies, products and organization in the innovating firm: What Adam. Smith tells us and Joseph Schumpeter doesn't. *Industrial and Corporate Change, 7*(3), 433–452.

Pavitt, K., & Patel, P. (1999). Global corporations and national systems of innovation: Who dominates whom. In D. Archibugi, J. Howells, & J. Michie (Eds.), *Innovation policy in a global economy* (pp. 94–119). Oxford University Press.

Peck, J., & Theodore, N. (2007). Variegated capitalism. *Progress in Human Geography, 31*(6), 731–772.

Pianta, M. (1995). Technology and growth in OECD countries, 1970–1990. *Cambridge Journal of Economics, 19*, 175–187.

Pierson, P. (2004). *Politics in time: History, institutions, and social analysis*. Princeton University Press.

Piore, M., & Sabel, C. (1984). *The second industrial divide: Prospects for prosperity*. Basic Books.

Pontusson, J. (2005). *Inequality and prosperity: Social Europe vs. liberal America*. Cornell University Press.

Porter, M. E. (1990). *The competitive advantage of nations*. McMillan Press Ltd..

Powell, W. W. (1991). Expanding the scope of institutional analysis. In W. W. Powell & P. J. DiMaggio (Eds.), *The new institutionalism in organizational analysis* (pp. 183–203). University of Chicago Press.

Ruggie, J. G. (1993). Territoriality and beyond: Problematizing modernity in international relations. *International Organization, 47*(1), 139–174.

Schmidt, V. A. (2007). *Bringing the state back into the varieties of capitalism and discourse back into the explanation of change*. Harvard University, Mind de Gunzburg Center for European Studies, Working Paper 152.

Schneider, M. R., & Paunescu, M. (2012). Changing varieties of capitalism and revealed comparative advantages from 1990 to 2005: A test of the Hall and Sokice claims. *Socio-Economic Review, 10*(4), 731–753.

Schoenberger, E. (1987). Technological and organizational change in automobile production: Spatial implications. *Regional Studies, 21*(3), 199–214.

Schoenberger, E. (1988). Multinational corporations and the new international division of labor: A critical appraisal. *International Regional Science Review, 11*(2), 105–119.

Scott, B., & Lodge, G. (Eds.). (1985). *US competitiveness in the world economy*. Harvard Business School Press.

Shepsle, K. (1989). Studying institutions: Some lessons from the rational choice approach. *Journal of Theoretical Politics, 1*(2), 131–147.

Shonfield, A. (1965). *Modern capitalism*. Oxford University Press.

Sorge, A., & Warner, M. (1986). *Comparative factory organisation: An Anglo-German comparison of manufacturing, management and manpower*. Gower Publishing Company.

Soskice, D. (1999). Divergent production regimes: Coordinated and uncoordinated market economies in the 1980s and 1990s. In H. Kitschelt, P. Lange, G. Marks, & J. Stephens (Eds.), *Continuity and change in contemporary capitalism* (pp. 101–134). Cambridge University Press.

Streeck, W. (2005). Industrial relations: From state weakness as strength to state weakness as weakness. Welfare corporatism and the private use of the public interest. In S. Greene &

W. E. Paterson (Eds.), *Governance in contemporary Germany: The semi-sovereign state revisited* (pp. 138–164). Cambridge University Press.

Streeck, W., & Thelen, K. (2005). Introduction: Institutional change in advanced capitalist economies. In W. Streeck & K. Thelen (Eds.), *Beyond continuity: Institutional change in advanced capitalist economies* (pp. 1–39). Oxford University Press.

Taylor, M. Z. (2004). Empirical evidence against varieties of capitalism's theory of technological innovation. *International Organization, 58*(3), 601–631.

Taylor, M. Z. (2016). *The politics of innovation: Why some countries are better than others at science and technology*. Oxford University Press.

Thelen, K. (1999). Historical institutionalism in comparative politics. *Annual Review of Political Science, 2*, 369–404.

Thelen, K. (2003). How institutions evolve: Insights from comparative-historical analysis. In J. Mahoney & D. Rueschemeyer (Eds.), *Comparative historical analysis in the social sciences* (pp. 208–240). Cambridge University Press.

Thelen, K. (2004). *How institutions evolve: The political economy of skills in Germany, Britain, the United States, and Japan*. Cambridge University Press.

Thelen, K. (2009). Institutional change in advanced political economies. *British Journal of Industrial Relations, 47*(3), 471–498.

Werle, R. (2012). Institutions and systems: Analysing technical innovation processes from an institutional perspective. In J. M. Bauer, A. Lang, & V. Schneider (Eds.), *Innovation policy and governance in high tech industries: The complexity of coordination* (pp. 23–48). Springer.

Whitley, R. (1992). *Business systems in East Asia: Firms, markets and societies*. Sage.

Whitley, R. (1999). *Divergent capitalisms: The social structuring and change of business systems*. Oxford University Press.

Whitley, R. (2005). How national are business systems? The role of states and complementary institutions in standardizing systems of economic coordination and control at the national level. In G. Morgan, R. Whitley, & E. Moen (Eds.), *Changing capitalism? Internationalism, institutional change, and systems of economic organization* (pp. 190–234). Oxford University Press.

Whitley, R. (2007). *Business systems and 0rganizational capabilities: The institutional structuring of competitive competences*. Oxford University Press.

Wojcik, D. (2002). The Länder are the building blocks of the German capital market. *Regional Studies, 36*, 877–895.

Zucker, L. G. (1983). Organizations as institutions. In S. B. Bacharach (Ed.), *Research in the sociology of organizations* (pp. 1–42). JA.

Chapter 4
Trade Fairs as "Organized Anarchies" and the Garbage Can Model of Organizational Choice

4.1 Introduction

Having reviewed the literature on international trade fairs (Chap. 2) and situated these events within the context of core debates in comparative political economy – specifically, the convergence-divergence debate and recent discussions regarding the dynamics of institutional change in relation to distinct capitalist varieties (Chap. 3) – this chapter turns to focus on the search processes of firms at international trade fairs. More specifically, this chapter describes the theoretical framework that informs the study's qualitative analysis of the search and interaction processes that take place among geographically – and, indeed, politically and economically – distant firms at these events (Chap. 5). This is an important step in establishing the micro-foundations underlying the study's main argument – namely, that international trade fairs represent a critical mechanism enabling the interaction of firms from different capitalist systems, and these inter-firm interactions can play a decisive role in the choices firms make with respect to their future pathways of technological specialization. Further, these pathways may, over time, have important implications for processes of cross-national convergence/divergence with respect to the technological specializations of firms.

In developing this final piece of the framework, this chapter draws important insights from three different yet complementary literatures: first, and most prominently, the literature on organizational decision-making, specifically the "garbage-can model of organizational choice" (Cohen et al. 1972); second, existing research on the technological search behaviours of firms; and, finally, relevant studies within the rapidly growing subfield of behavioural economics. Among these three literatures, I suggest that the garbage-can model is especially helpful in conceptualizing the nature of firms' search and interaction processes at trade fairs. A critical reason for this is that the garbage-can model draws special attention to the role of discourse and ideas in shaping both individual and organizational decision-making, both of

R. Gibson, *International Trade Fairs and Inter-Firm Knowledge Flows*,
https://doi.org/10.1007/978-3-031-20557-6_4

which are integral to the specific communication and information ecology that occurs at, or is stimulated by, trade fairs. Although the original formulation of the garbage-can model had universities in mind when describing organizations as "collections of choices looking for problems" and "solutions looking for issues to which they might be the answer" (Cohen et al., 1972, 2), I suggest that this characterization is also well suited to the complex organizational setting of international trade fairs.

The remainder of this chapter proceeds as follows. The next section reviews the literature on organizational decision-making, with a particular focus on the influential work of Cohen et al. (1972) – namely, the "garbage-can model of organizational choice". Through the course of my own empirical research, I have observed that the actions and interactions of firms at trade fairs closely follow the essential logic of the garbage-can model. Thus, in addition to outlining the core components of the garbage-can model, this chapter also places considerable emphasis on how this model can be used to characterize and investigate the dynamics of firm actions and interactions during international trade fairs. To complement the work of Cohen et al. (1972), the second and third parts of this chapter turn to other literatures, both of which help to deepen our understanding of the nature of firms' search and interaction processes at trade fairs. First, lessons are drawn from existing research on organizational or technological search processes and, second, from studies in behavioural economics. With respect to the latter body of work, I am particularly interested in how the ideas of behavioural economics contribute to the discussion of how organizations and firms process information and make decisions. Focusing on the micro-level, this chapter represents the final part of the study's overarching framework for analysis.

4.2 Organizational Decision-Making: The Garbage-Can Model[1]

In applying the garbage-can model to the search processes of firms at trade fairs, this study recognizes that trade fairs can be viewed both as events and as organizational forms. They may be viewed as events, for example, when the emphasis is placed on the temporary nature of meeting with peers in a given industrial or technological field. Yet, when one considers their role in defining the rules, shared goals, specific circumstances, and settings for interaction and knowledge circulation, they can also easily be viewed as an organizational form. This study adopts this latter focus. Although trade fairs are different from organizational forms like markets, hierarchies (firms), networks, and projects, they share some common features. Trade fairs are less focused on immediate transactions; however, knowledge circulation and exchange are a fundamental part of the actions and interactions that take

[1] This section of the chapter is based on Bathelt and Gibson (2015).

place. Similar to firms, networks, and projects, trade fairs are defined by membership or participation. Trade fair organizers create the rules of interaction and exchange; they also have the authority to determine access to trade fairs. Yet, the nature of interaction at trade fairs is temporary. In this respect, it may be more akin to the type of interaction that is found in projects and, at times, can even resemble one-off market relations. Important interactions nevertheless take place, especially within the context of ongoing producer-user relations that may become part of networks (Bathelt & Gibson, 2015). Despite some apparent commonalities, this study suggests that the nature of interaction at trade fairs is less predictable and less systematic than many other organizational forms. As such, it is an organizational form that better fits the basic characteristics of an "organized anarchy", which will be described next.

4.2.1 "Organized Anarchies"

Originally developed by Cohen et al. (1972) to analyze decision-making processes in the university setting, the garbage-can model of organizational choice may be viewed as part of a larger critique of rational-choice decision-making theories. The garbage-can model is distinct from most other organizational choice theories in that it highlights the non-linear and often ambiguous dynamics of decision-making in organizations that share the characteristic features of "organized anarchies." Organized anarchies are defined by the following three core properties (Cohen et al., 1972): (i) problematic preferences, (ii) unclear technologies or processes, and (iii) fluid participation. As discussed below, these properties go a long way in capturing the dynamics of the trade fair – an organizational form that is both partially open in nature as well as defined by a specific kind of temporality. In applying the garbage-can model to international trade fairs, this study suggests that these events represent an organizational form that is similar to other complex organizational forms to which the model has been applied in the past. Such examples include: networks (Lomi & Harrison, 2012), universities (Cohen et al., 1972), and governments (Kingdon, 2003).

With respect to the first property, "problematic preferences", it is suggested that individuals generally do not define their preferences very clearly. This does not prevent action or decision-making from taking place. As Lindblom (1959) observes, not only does action occur in the absence of well-defined goals, it is also frequently enabled by glossing over what one aims to achieve. Moreover, even when individuals do define their preferences fairly precisely, such preferences tend to come into conflict with others. This means that different decision-makers within an organization will often disagree about the overarching goals of their organization; it also means that a given individual will change or rethink their own preferences over time. As such, the organization may be aptly described as "a loose collection of ideas [rather than] a coherent structure; it discovers preferences through action more than it acts on the basis of preferences" (Cohen et al., 1972, 1).

International trade fairs have been described as representing the intersection of focused communities (Bathelt & Schuldt, 2008). Yet, the individual participants at these events (in this case, the trade fair exhibitors) differ substantially in how they evaluate new trends and in what they consider important and why. Indeed, it may be argued that much of the contingency of economic action relates to the fact that economic agents often make fundamentally different choices under the same or similar circumstances. Developing a better understanding of the nature of such heterogeneities is important in establishing the micro-foundations of economic behaviour. As for problematic preferences, these are a prominent feature of trade fairs because participants at these events often hold diverging views about the objectives of the event or the direction of change in the industry. It is also quite common for trade fair participants to rethink their own goals over time. As such, trade fairs are very much an organization that embodies a "loose collection of ideas", where preferences are identified and defined through the course of action itself, rather than actions being based on clearly defined preferences or goals (Cohen et al., 1972, 1).

With respect to "unclear technologies or processes", the participants of an organized anarchy rarely have a clear idea of the organization's processes. While participants may have a solid grasp of their own jobs or position within the organization, they have only a basic or partial understanding of how their jobs fit into the broader picture of the organization (Kingdon, 2003). On the rare occasions when individuals agree on the organization's objectives, identifying the exact technologies needed to realize such goals often remains difficult (Fardal & Sørnes, 2008). As such, organized anarchies typically operate according to a method of "trial-and-error", by learning from past errors or missteps, and through "pragmatic inventions of necessity" (Cohen et al., 1972, 1).

In the context of an international trade fair, participants are regularly confronted by a high degree of technological and market uncertainty. They rarely have precise ideas about how best to choose a technology and, as such, tend to proceed experimentally at first, making decisions with only partial knowledge of the consequences of their actions. As described by Cohen et al. (1972, 2), "[p]articipants arrive at an interpretation of what they are doing and what they have done while in the process of doing it". As will be discussed in greater detail in Chap. 5, many trade fair exhibitors do not come to the event with a preconceived notion of what they expect to discover or achieve during the trade fair. Instead, many participants suggest that they will come across something that is new or relevant by wandering around the exhibition hall and waiting for something to "catch their eye".

"Fluid participation" is the final property attributed to organized anarchies. It refers to the fact that participants tend to "come and go" in an organization and will devote varying amounts of time and energy to different organizational issues (Cohen et al., 1972). The boundaries of such organizations are, therefore, unclear and characterized by considerable instability. The central decision-makers in such organizations may also change on a regular or ongoing basis. This notion of fluid participation in an organizational setting is reflected in the context of the trade fair as well. While firms within a specific industry often attend and exhibit at the same trade fairs over multiple years, the representatives of these firms are not always the same. This is particularly the case with large multi-national firms, many of which have a sizeable staff to devote to marketing and promotional activities such as trade fair representation. As Kingdon (2003, 84) observes in the context of the federal government, "[w]ho shows up for or is invited to a given critical meeting, and their degree of activity at the meeting, for instance, turn out to make a tremendous difference". Certainly, the same could be said of exhibiting firms at international trade fairs, where pivotal meetings are often held or planned. Such meetings can lead to a new important business deal, a key insight into new technological innovations, or the start of a new partnership or collaboration that could influence the future direction of a firm.

To summarize, the ambiguity of organizational decision-making in organizations that share the defining features of the organized anarchy is due to the absence of stable and clearly defined goals (i.e., problematic preferences), the indeterminacy of knowledge and the appropriate methods needed to realize organizational objectives (i.e., unclear technologies), and the sporadic attention, involvement, and presence of participants in decision-making (i.e., fluid participation). According to Cohen et al. (1972, 1), organized anarchies are not a unique type of organization; the three properties just described are "characteristic of any organization in part – part of the time." In these settings, decision-making rarely follows rational or systematic channels. Instead, organizational choices are typically a result of the interrelations among multiple independent process streams that flow through the system. These are: (i) a stream of problems, (ii) a stream of solutions, and (iii) a stream of participants (Cohen et al., 1972; see also, Kingdon, 2003).[2] A brief description of each follows below.

4.2.2 *The Process Streams*

From the garbage-can model perspective, the linkages between means and ends or actions and outcomes are neither straightforward nor predictable. As March and Olsen observe (1984, 740), the "connection between problems and solutions may be less dominated by a logic of causal linkages between means and ends than by less problematic temporal linkages of simultaneity". In other words, bringing together the various critical ingredients in the same place at the same time is, perhaps, the

[2] In the original garbage-can model, Cohen et al. (1972) identify four streams or flows of elements – a stream of problems, solutions, participants, and choice opportunities. Kingdon (2003) modifies this set of processes for his model by substituting "solutions" with "policies" and "participants" with "politics". He also discusses the fourth stream - "choice opportunities" or "policy windows", in Kingdon's formulation – separately from the first three streams. Despite these modifications, Kingdon subscribes to the original model's core assumption that the process streams are largely independent of one another – or "loosely coupled" (Cohen et al., 1972). I find Kingdon's separation of policy windows from the other three streams useful in conceptualizing the role of time and chance in the interactions among the different streams, but I retain Cohen et al. (1972) original categories of streams – namely, problems, solutions, and participants.

most powerful condition for organizational decision-making. From the garbage-can model perspective, these critical ingredients take the form of three largely independent streams of processes: (i) problems, (ii) participants, and (iii) solutions.

While the timeline of impact is different, a similar argument has been made within the context of research on industrial clusters, where the act of just "being there" (Gertler, 1995, 2003) is vital in driving processes of localized learning and knowledge exchange. Although the idea of "temporal linkages of simultaneity" may be applied to a wide range of empirical settings or problems, this idea seems particularly relevant in the context of trade fairs. Indeed, trade fairs are organized according to clearly delineated temporal boundaries, usually ranging from 2 to 7 days depending on the trade fair. In a given location and over a specified period of time, the trade fair brings together what Cohen et al. (1972) refer to as three major process streams – problems, solutions, and participants.

The *problems stream* refers to the processes by which a particular issue or event – be it social, political, or economic in nature – comes to be recognized as "a problem" by the relevant group of actors. Problems can originate both inside and outside of an organization. Objective indicators, such as a revolutionary new process design, may draw widespread attention to an emerging problem in an industry; yet, in most cases, problems are not entirely self-evident. As such, determining what constitutes an actual "problem" at any point in time is often subject to a large degree of individual interpretation (Kingdon, 2003). Decision-making within the problems stream, therefore, typically requires ongoing deliberation among the participating actors. And, defining such a problem can often involve a deeply political process in which a contest of ideas, values, and norms will give legitimacy to some issues and not others.

As for the *solutions stream*, the garbage-can model challenges conventional wisdom regarding the causal direction of the problem–solution relationship in that it views actors not as "problem-solvers" but, rather, as "solution-marketers" (Fardal & Sørnes, 2008). In the context of the present research, a "solution" refers to a new product or process innovation presented or discovered by a firm at an international trade fair. And, since the garbage-can model treats the "offering of solutions" as a key part of organizational action (Fardal & Sørnes, 2008), leading international trade fairs are expected to provide a valuable test environment for the analysis of such behaviour. Trade fairs can, thus, be usefully conceptualized as an organizational form that accommodates a "collection of choices looking for problems, issues and feelings looking for decision situations in which they might be aired, solutions looking for issues to which they might answer, and decision-makers looking for work" (Cohen et al., 1972, 1).

Finally, with respect to the *participants stream*, organizational membership is described as both unstructured and transient. Not only do participants come and go as they please, they also bring different skills or strengths, as well as different amounts of interest or energy to the organization. External factors, such as time, are also important in determining which projects participants will focus on and how much time they will spend on organizational decision-making. Even if the focus is solely on exhibitors, as it is in this case, the participants at trade fairs are incredibly diverse. Take, for example, participants' position within the exhibiting firm. Such participants can range from the firm's marketing representative to its engineer to its president or CEO. As such, trade fair participants vary substantially in terms of the skills, knowledge, experience, and interest they bring to the event.

Although described as largely independent of one another – "each with a life of its own" (Kingdon, 2003, 86) – the different streams come together at key junctures, and this "coupling" is what leads to a significant decision or change being made. In the original formulation (Cohen et al., 1972), these important junctures were referred to as "choice opportunities", whereas later prominent versions of the model (Kingdon, 2003) have used the term "windows of opportunity". As an organizational form, trade fairs may be usefully conceived as "organized anarchies" comprising the three core properties of problematic preferences, unclear technologies or processes, and fluid participation.[3] Yet, at the same time, the nature of many trade fairs facilitates the very "windows" or "choice opportunities" required to achieve fundamental decision-making in such settings. In this respect, the specific features of an organizational structure can play an important role in shaping garbage-can decision-making processes and outcomes. As Cohen et al. (1972, 4) explain, "[e]lements of organizational structure influence outcomes of a garbage can decision making process by: (i) affecting the time pattern of the arrival of problems, choices, solutions, or decision makers; (ii) determining the allocation of energy by potential participants in the decision; and (iii) establishing linkages among the various streams." In relation to trade fairs, it is the organizational structure of the trade fair itself that critically guides the actions and interactions of firms operating within it. In this context, the "windows" or "choice opportunities" might refer to the establishment of a potential business partner or an alliance, the achievement of an organizational agreement on the viability of a new technological innovation, or the time, money, and responsibilities that are to be allocated to a particular endeavour.

4.2.3 The Ongoing Relevance of the Garbage-Can Model

Over the past four decades, the garbage-can model has been modified and used to analyze decision-making processes in a wide range of organizational settings. It has been used, for instance, to understand decision-making dynamics in the context of military operations (Bromiley, 1986; March & Weissinger-Baylon, 1986), public

[3] In characterizing the trade fair as an "organized anarchy", I am not suggesting that all such events fit this description perfectly. Indeed, there are various types of trade fairs, organized in different ways and directed towards different purposes (for an extended discussion, see Bathelt et al., 2014). The concept of the organized anarchy nevertheless captures key elements of the trade fairs analyzed in this study, and probably certain aspects of most types of trade fairs at least some of the time. As Cohen et al. (1972, 1) are careful to note, the "properties of organized anarchy [...] are characteristic of any organization in part, part of the time. They are particularly conspicuous in public, educational, and illegitimate organizations. A theory of organized anarchy will describe a portion of almost any organization's activities, but will not describe all of them."

policy-making and agenda-setting in the U.S. federal government (Kingdon, 2003), foreign policy-making (Newmann, 1998), and educational planning in organizations (Clark et al., 1980). More recently, the garbage-can model has been applied to the Social Security reform initiatives of the Bush administration (Weiner, 2007), the failure of the Kyoto Protocol (Honda, 2007), and strategic decision-making in information technology systems (Fardal & Sørnes, 2008). Empirical applications of the model include an equally diverse array of methodological approaches, from laboratory experiments to ethnographic research to equation-based and agent-based computer simulations (Lomi & Harrison, 2012).

A recent edited volume celebrating the 40th anniversary of the original work of Cohen et al. (1972) offers further evidence not only of the widespread influence of the model, but also of its ongoing relevance. As the editors of the book observe, the original paper is "one of the most influential and sustained attempts to represent organizational decision-making processes in a way that accounts for generally recognized but hard to accept features of organizational life" (Lomi & Harrison, 2012, 3). Yet, despite its demonstrated applicability to a variety of research topics, the garbage-can model has not been used to investigate the search and information processes of firms at international trade fairs (for an exception, see Bathelt & Gibson, 2015). Given the model's emphasis on the diffusion of new ideas or "solutions" through seemingly random or non-linear processes, this chapter argues that the garbage-can model is an extremely useful framework for conceptualizing the search and information processes of firms at international trade fairs. Before using the garbage-can model to inform the empirical analysis presented in Chap. 5, this chapter first explores two other literatures that offer further insights regarding the microfoundations of inter-firm interaction at international trade fairs. I begin with an overview of the existing literature on organizational and technological search.

4.3 The Organizational Search Literature

Technological search has long been considered a pivotal part of organizational learning and innovation processes (Nelson & Winter, 1982; Laursen, 2012; Lopez-Vega et al., 2016). It is through search that firms try to solve existing problems and through search that they find ways to adapt in a complex and continuously changing environment (Huber, 1991; Fleming & Sorenson, 2001; Argote & Miron-Spektor, 2011; Wilson & Joseph, 2015). While there is quite a substantial literature on firm-level search processes,[4] much of the research in this area centres on one of two search dimensions. The first dimension is interested in *where* to search with respect to the location of available solutions (Lopez-Vega et al., 2016). This search dimension is referred to in the literature as: *exploitative* vs. *explorative* search (March, 1991; Helfat, 1994; Stuart & Podolny, 1996; Fleming & Sorenson, 2004; Nerkar &

[4] For a more comprehensive review of this literature, see Laursen (2012).

Roberts, 2004; Laursen, 2010; Wilson & Joseph, 2015).[5] The second search dimension focuses on *how* to search. This dimension is described in the literature as: *experimental* (or "directional) vs. *cognitive* (or "heuristic") search (Nickerson & Zenger, 2004; Gavetti & Rivkin, 2007; Lopez-Vega, et al. 2016).

Before describing each dimension in greater detail, it is important to note that existing research on organizational search predominantly employs a unidimensional construct for the analysis of these processes (Lopez-Vega et al., 2016). Indeed, only a few studies pay attention to both the *where* and *how* dimensions of search and how these two interact – for example, how the "location" of solutions (e.g., closely related to the firm's existing knowledge base) interacts with search heuristics (Lopez-Vega et al., 2016). While not a primary goal of the present study, the analysis of firms' search processes at trade fairs (presented in Chap. 5) contributes to the literature on organizational search in the sense that it explores both "where" and "how" firms search at these events. First, however, it is necessary to establish a clearer understanding of the two dominant search dimensions discussed in the search literature: (i) *exploitative vs. explorative search*, and (ii) *experiential vs. cognitive search*. To this end, a brief description of each follows below.

4.3.1 Exploitative vs. Explorative Search

Before discussing exploitative and explorative search, it is important to stress that the focus here is on technological searches; as such, the empirics and discussion presented below emphasize this particular aspect of search as opposed to others. When firms are engaged in *exploitative* search, it means that they search for solutions that closely relate to their pre-existing bases of knowledge (Helfat, 1994; Martin & Mitchell, 1998; Gavetti et al., 2012; Wilson & Joseph, 2015). When firms are engaged in *explorative* search, on the other hand, they are involved in "a conscious effort to move away from current organizational routines and knowledge bases" for the purpose of gaining access to new external knowledge (Katila & Ahuja, 2002, 1184; March, 1991; Rosenkopf & Nerkar, 2001; Wilson & Joseph, 2015). Empirical applications of these concepts help to further clarify the nature of exploitative and explorative search processes (Laursen, 2012). The work of Katila and Ahuja (2002, 1183), for instance, extends the exploitative-explorative dimension by adding two new search variables: *search depth* and *search scope*. The first relates to exploitative search and refers to "how deeply a firm reuses its existing knowledge" (Katila & Ahuja, 2002, 1183). The second is associated with explorative search and refers to "how widely a firm explores new knowledge" (Katila & Ahuja, 2002: 1183). Using patent data as a measure of search, Katila and Ahuja

[5] The majority of this literature uses the terms "local" versus "non-local" or "distant", instead of "exploitative" versus "explorative". Yet, in an effort to avoid confusion with geographical interpretations of the terms "local", "non-local", and "distant" found in Economic Geography, for instance, this study will use the terms "exploitative" and "explorative" throughout the discussion.

(2002, 1187) capture *search depth* by calculating the "average number of times a firm repeatedly used the citations in the patents it applied for". To capture *search scope*, the same study looks at the share of new citations in a given year's citations – that is, those that "could not be found in the previous five years' list of patents and citations by that firm" (Katila & Ahuja, 2002, 1187). Other studies have employed similar measures of exploitative vs. explorative search (see, for example, Benner & Tushman, 2002; Laursen et al., 2010).

The bulk of the literature finds that firms mainly engage in exploitative search processes, rather than explorative ones, and that this type of search is typically the most efficient due to its reliability and relatively low costs (Pavitt, 1988; Cantwell, 1989; Helfat, 1994; Stuart & Podolny, 1996; Tripsas & Gavetti, 2000; Laursen, 2012). Explorative search is generally seen as a more difficult undertaking because firms cannot rely on their own experience or established routines for guidance (Nelson & Winter, 1982; Wilson & Joseph, 2015). As Pavitt (1988, 130 [emphasis original]) explains,

> the search processes of industrial firms to improve their technology is *not* likely to be one where they survey the whole stock of technological knowledge before making their technical choices. Given its highly differentiated nature, firms will instead seek to improve and to diversify their technology by searching in zones that enable them to use and to build upon their existing technological base.

The assumptions of "bounded rationality" (Simon, 1955, 1987b) are significant here and have formed an important basis for discussion around organizational search processes (Cyert & March, 1963; Gavetti & Levinthal, 2000). Indeed, at the heart of this notion is a recognition of the "numerous cognitive limitations" (Simon, 1987b, 222) of economic actors and the tendency towards "satisficing" rather than utility maximization in the decision-making realm.[6] As will be discussed at greater length in the next section of this chapter, the notion of bounded rationality is a core concept in behavioural economics research. The assumptions of bounded rationality also help to inform this study's exploration of the search and information processes of firms at trade fairs.

Although the empirical literature suggests that firms mainly engage in exploitative searches, studies drawing on March's (1991) influential work point to the pitfalls of relying too heavily on existing routines and past solutions – or what some have called the "local search trap" (Laursen, 2012, 1195; see also: Levinthal & March, 1993; Stuart & Podolny, 1996; Gavetti & Levinthal, 2000; Tripsas & Gavetti, 2000; Postrel, 2002; Levinthal & Rerup, 2006). Simply put, if a firm becomes too dependent on exploitative search, it may have difficulty innovating or coming up with novel solutions to new complex problems. This is because the knowledge that

[6] As originally proposed by Simon (1955, 1956) the concept of "satisficing" is at the core of decision-making under bounded rationality, as it describes the processes by which actors examine and evaluate alternatives. In contrast to the tenets of classical economics, it suggests that actors rarely seek out optimal solutions but, rather, solutions that are "good enough". The difference, as Simon (1987b: 244) explains, is that optimizing involves "looking for the sharpest needle in the haystack", whereas satisficing involves "looking for a needle that is sharp enough to sew with".

is in the vicinity of a firm's pre-established knowledge bases may not always supply the necessary variety to solve problems, or the opportunities needed for combination and recombination of knowledge (Rosenkopf & Nekar, 2001; Fleming & Sorenson, 2004; Laursen, 2012). Heavy dependence on exploitative search can also lead to organizational myopia (March, 1991; Levinthal & March, 1993) – as, for example, when firms systematically fail to recognize superior solutions because such solutions are located in distant knowledge communities or fields (Laursen, 2012).

Acknowledging the need to access solutions that are beyond the firm's existing knowledge base, some studies have focused on "boundary spanning" as a mechanism by which to overcome the imbalance between exploitative and explorative search processes (Lopez-Vega et al., 2016). As noted in the work of Lopez-Vega et al. (2016, 126; see also, Rosenkopf & Nekar, 2001), "[a]lthough exploring within the boundaries of the firm's technological domain may satisfy a specific technology need, boundary spanning involving a distant technological domain helps identify new ways to solve problems". In this context, some research has pointed to the role of "innovation intermediaries" in bridging the exploitative-explorative search space (Lopez-Vega et al., 2016). Innovation intermediaries have been defined as "[a]n organization or body that acts as an agent or broker in any aspect of the innovation process between two or more parties" (Howells, 2006, 720). They are "organizations or groups within an organization that work to enable innovation, either directly by enabling the innovativeness of one or more firms, or indirectly by enhancing the innovative capacity of regions, nations, or sectors" (Dalziel, 2010, 3–4). Examples of innovation intermediaries found in the literature include: trade associations, chambers of commerce, economic development agencies, science, technology, and business parks, business incubators, research consortia and networks, research institutes, industrial liaison offices, and innovation centres (Phillimore & Joseph, 2003, 751; Eggink, 2013, 9). Innovation intermediaries can bridge the exploitative-explorative search dimension by enabling different specialized processes or activities, such as organized innovation contests that bring together distinct knowledge communities (Spradlin, 2012; Lopez-Vega et al., 2016). This study suggests that leading international trade fairs might also be added to the above list of innovation intermediaries.

The above discussion regarding the need for balance between exploitative and explorative search processes parallels recent discussions in the literature on the competitiveness of industrial clusters and the conditions required for cluster growth. Recent research on the importance of both "local buzz" and "global pipelines" (Bathelt et al., 2004), for example, seems particularly relevant. The basic argument here is that the competitiveness of clusters largely depends on the presence of both high quality "local buzz" and a well-developed system of "pipelines" or extra-local knowledge sources, which link the cluster to the outside world (Bathelt et al., 2004; Bathelt, 2005). Without engaging fully with this argument, for present purposes it is important to note that international trade fairs have been positioned in this literature as important vehicles through which global pipelines may be established (Maskell et al., 2006). Similarly, international trade fairs – acting as "innovation intermediaries" – might play an important role in balancing the emphasis firms tend to place on exploitative versus explorative searches.

4.3.2 Experiential vs. Cognitive Search

As mentioned above, the second search dimension, which has received relatively less attention in the literature, focuses on *how* searches are conducted. In this body of work, special attention is paid to the search heuristics firms' use to make decisions about solutions that may be relevant to them, as well as the specific processes involved (Gavetti & Levinthal, 2000; Gavetti & Rivkin, 2007; Lopez-Vega et al., 2016). The two main types of search in this literature are: *experiential* and *cognitive*.

Experiential or experience-based search refers to the search for solutions through a process of "trial-and-error", where firms experiment among the various alternatives available to them, receive feedback, and respond accordingly to this feedback (see, for example, March & Simon, 1958; Cyert & March, 1963; Nelson & Winter, 1982; Hsieh et al., 2007). Experiential search has been described as "backward-looking" (Gavetti & Levinthal, 2000) in that it relies on direct feedback from the environment to guide future action and apply solutions. This type of search is characterized by a process of "learning-by-doing" (Pisano, 1994), whereby the firm gathers insight into what solutions may be feasible or desirable while embedded in the relevant learning context. Experiential search has also been described as the "on-line" evaluation of alternatives because "experiential processes inherently require at least partial implementation of an alternative in order to evaluate its efficacy" (Gavetti & Levinthal, 2000, 115). In other words, a choice is made, the consequences of that choice are felt, and revisions to the initial choice or action can subsequently take place (Levitt & March, 1988; Gavetti & Levinthal, 2000). "Routine development" is a good example of experiential search processes (Nelson & Winter, 1982). Recalling the discussion of the garbage-can model above, experiential search processes also correspond to one of the defining features of organized anarchies – namely, "unclear technologies or processes".

In contrast to experiential search, cognitive search has been described as "forward-looking" (Gavetti & Levinthal, 2000); it relies much more heavily on representations and abstractions in the quest for solutions (Gavetti, 2012). This type of search involves "learning-*before*-doing" (Pisano, 1994) and the "off-line" evaluation of alternatives, which means that firms do not need to engage in an activity in order to evaluate it (Gavetti & Levinthal, 2000; Lopez-Vega et al., 2016). Cognitive search does not incorporate the direct feedback of environmental stimuli into action; rather, it enables the "feed-forward of information into existing or nascent representations" that can be examined and later enacted (Lopez-Vega et al., 2016, 127). These representations play an important role in the development of different heuristics. Learning vicariously from other organizations is, for example, a type of heuristic search (Cyert & March, 1963; Yang et al., 2010). As Yang et al. (2010, 7) observe, "[o]rganizations learn vicariously by observing the behavior and associated performance outcomes of other organizations and then modeling or imitating behaviors that seem successful and avoiding behaviors that seem unsuccessful". This heuristic seems particularly pertinent to international trade fairs, as these events offer a unique opportunity for firms to learn from and obtain an overview of what other firms are doing. During the trade fair, firms have a chance to watch how competitors interact with customers, suppliers, and other relevant economic actors. They also have an opportunity to witness, first-hand, the display of new technological innovations in products or processes, and compare their own performance with that of their competitors. While the literature on cognitive search identifies a variety of different types of search heuristics, it does not tell us much about which heuristics or pathways are dominant and why. The empirical analysis of firms' search and information processes, presented in Chap. 5, contributes further insights into these aspects of cognitive search.

4.3.3 Organizational Search and International Trade Fairs

In light of the above discussion regarding organizational or technological search, Chap. 5 takes stock of both "where" and "how" exhibiting firms search during trade fairs. When it comes to the experiential-cognitive dimension (i.e., *how* firms search), for example, I expect to find evidence of greater engagement in experiential than cognitive search processes. This is due, in part, to the opportunities trade fairs provide for firms to engage in multi-sensory explorations of potential solutions – for example, seeing and touching different artifacts at the exhibit booths. Processes of "learning-by-doing" and "trial-and-error", characteristic of experiential search, also resonate strongly with the trade fair experience. On the question of exploitative versus explorative search, the expectations may be a bit less clear. Although trade fairs represent key vehicles through which explorative search may be accomplished, during these events, firms may still display a stronger tendency toward search that is guided by deeply-rooted knowledge bases and pre-established routines and practices. This apparent tension corresponds to discussions about the "paradox of globalization" (Bathelt & Glückler, 2011), which may be summarized as follows: "Although the exchange of goods and knowledge across regional, national, and cultural boundaries is becoming easier thanks to modern technologies, we are faced with the paradox that global economic success still depends on the utilization of local and/or regional resources" (Bathelt & Glückler, 2011, 131).

In addition to using the search literature to shed light on the search processes of firms at international trade fairs, the study of trade fairs itself can also play a valuable role in extending theoretical and empirical research on technological search. From a purely methodological standpoint, trade fairs offer a valuable site in which to explore a number of open questions regarding, for instance, the search heuristics used by firms and the nature of the interactions between different search dimensions (i.e., the exploitative-explorative dimension, and the experiential-cognitive dimension). As previously mentioned, it may also be useful to consider international trade fairs along the lines of "innovation intermediaries", exploring the specific ways in which these events help balance the exploitative-explorative search dimension. Finally, with respect to moving beyond the one-dimensional frameworks commonly used to analyze organizational search, the present study may offer some insight into how to capture the multi-dimensional facets of search. While not its main objective, this study builds on the existing search literature by contributing new empirically-grounded insights into the nature of search processes as well as the search heuristics firms' use within the context of international trade fairs.

A basic assumption in the organizational search literature is that firms engage in search processes in order to discover solutions to problems. This search for solutions can assume two different forms: (i) the direct search for solutions, or (ii) the search for information regarding solutions that have already been identified by the firm. Although much of the literature fails to provide a clear conceptual distinction between the two types of solution-oriented searches, the first is generally treated as more a question of "discovery or identification", whereas the latter is more about "investigation" (Huber, 1991, 99). While organizational search is often carried out for the purpose of learning about innovation and discovering "solutions", my research suggests that the search objectives of firms are often more ambiguous than might be expected and that the key actors themselves (in this case, trade fair exhibitors) often do not have a clear sense of what they are searching for until they come across it. Thus, drawing on insights from the garbage-can model regarding decision-making in "organized anarchies", this study contributes to discussions in the organizational search literature by challenging the traditional problem-solution dynamic and documenting the non-linear, or garbage-can-like, processes that seem to be characteristic of searches at trade fairs. As will be developed more fully in the empirical part of the book (Chaps. 5 and 6), this finding is central to the argument that inter-firm interaction at trade fairs does not lead automatically to technological convergence but, instead, may be linked to patterns of continued divergence.

In addition to the garbage-can model and the search literature discussed above, empirical findings from behavioural economics shed further light on the complexities and contingencies associated with individual judgment and choice, which, in turn, also problematize rational actor interpretations of decision-making. A brief discussion of key ideas from this literature and how they contribute to the analysis of inter-firm interaction at trade fairs follows below.

4.4 Studies in Behavioural Economics

As an independent subdiscipline, "behavioral economics" has been described as "the attempt to increase the explanatory and predictive power of economic theory by providing it with more psychologically plausible foundations" (Angner & Loewenstein, 2012, 642; see also, Camerer & Loewenstein, 2004). Among other things, this involves appreciating and exploring "the implications of the limits of rationality" (Ho et al., 2006, 307). Behavioural economics is deliberately interdisciplinary in nature, borrowing from neighbouring fields such as psychology and sociology. In this way, it is actually "an umbrella of approaches that seek to extend the standard economics framework to account for relevant features of human behavior that are absent in the standard economics framework" (Diamond & Vartiainen,

2007, 1). In recent years, the core principles and models of behavioural economics have been applied to a growing number of research areas and problems. Yet, for the purposes of this study, I am primarily interested in how behavioural economics can help to illuminate the actions and interactions of firms in a specific decision-making setting – international trade fairs. As such, what follows is a cursory overview of some of the key ideas and themes from the behavioural economics literature that have a particular bearing on the research at hand.[7]

4.4.1 *Bounded Rationality*

A first key concept that has been influential in behavioural economics, as well as in other disciplines, is that of bounded rationality. As briefly mentioned above, the term "bounded rationality" was initially proposed by Simon (1987a, 222; see also Simon, 1955) to signify "the whole range of limitations on human knowledge and human computation that prevent economic actors in the real world from behaving in ways that approximate the predictions of classical and neoclassical theory". In other words, the "boundedly rational" actor is one with "limited computational ability and, perhaps, imperfectly defined objectives, attempting to cope with an often complex decision environment" (Starmer, 2004, 126). While the aim of neoclassical choice theory is not to elucidate the precise mechanisms by which actors make decisions, but rather to serve as "an apparatus for predicting choice", Simon (1987b: 267) argues that theories of bounded rationality have greater ambitions in that they try to "capture the actual process of decision as well as the substance of the final decision itself".

When it comes to extending the bounded rationality perspective into the realm of organizational or firm behaviour, the influential works of March and Simon (1958), *Organizations*, and Cyert and March (1963), *A Behavioral Theory of the Firm*, are the most frequently-cited starting points.[8] These works, and the theory of bounded rationality, in particular, have had a significant impact on organizational theories in a variety of disciplines and subfields, including institutional theories (Meyer & Rowan, 1977; DiMaggio & Powell, 1983), organizational economics (Williamson, 1985; Radner, 1996), population ecology (Miner et al., 1990; Barnett & Hansen, 1996), as well as political science and public administration (Kingdon, 2003). Of course, included in this body of work is also the garbage-can model of organizational choice discussed in greater detail above (Cohen et al., 1972).

Firms participating at international trade fairs are confronted with vast amounts of information. This information can be related to trends in industries, markets,

[7] For a more detailed discussion and historical overview of the development of behavioural economics as an independent subdiscipline, see: Camerer et al. (2003), Camerer and Loewenstein (2004), Angner and Loewenstein (2012).

[8] For a useful review of the impact of these works on different fields and programs of research, see Argote and Greve (2007).

technologies or innovations; it can also be related to customers, competitors, networks, as well as potential partners or alliances. Moreover, the information at trade fairs is circulated and received in a multitude of ways: in the form of news, strategic information, first-hand, observational or sensory experiences, rumours or gossip, recommendations, as well as both hard data and speculations about an industry or firm (Schuldt & Bathelt, 2011). Acknowledging the bounded rationality of firms in such an information-rich environment is important in understanding the nature and dynamics of inter-firm interaction at trade fairs. Engaging in search processes that take into account the entire range of information inputs identified above is impractical if not impossible, especially when one considers that much of this information is dynamic and received in the moment. Moreover, exhibiting firms are constrained in terms of the amount of time and resources they can actually devote to search activities. Indeed, exhibiting firms are preoccupied with many other activities during the trade fair, including representing their own products or services through demonstrations and talks, meeting with new and existing customers, suppliers, and distributors, as well as participating in workshops, innovation contests, and so on. In the face of such complex realities, firms often need to rely on mental shortcuts or heuristics in order to make decisions.

4.4.2 Heuristics and Biases

Closely related to bounded rationality are the heuristic shortcuts actors employ to deal with large amounts of information. Based on the pioneering work of Tversky and Kahneman (1974, see also Tversky & Kahneman, 1983), studies in behavioural economics draw attention to the role of heuristics in decision-making and how the use of heuristic mechanisms or shortcuts can lead to judgments or choices that challenge the principles of probability (Camerer et al., 2004; Kahneman & Frederick, 2002). Such heuristic shortcuts enable more efficient decision-making in situations where a comprehensive assessment of the full array of possibilities is unmanageable or unrealistic due to a lack of time and cognitive resources (Kahneman et al., 1986; Gigerenzer & Todd, 1999; Katapodi et al., 2005). A variety of different heuristics have been explored in the context of behavioural economics research, including, most notably, the *availability* heuristic, the *representativeness* heuristic, as well as the *anchoring-and-adjustment* heuristic (Tversky & Kahneman, 1974; Choi & Pritchard, 2003).[9] A brief description of each is provided below.

---

[9] For a detailed description and discussion of each of these heuristics and their associated biases, see Tversky and Kahneman (1974). There is also a vast amount of empirical research focused on one or more of these heuristics and the many biases they bring about. For recent work on the availability heuristic and related biases, see: Schwarz and Vaughn (2002), Choi and Pritchard (2003). On representativeness, see: Khaneman and Frederick (2002), Bar-Hillel and Neter (2002). And for recent studies on anchoring and adjustment, see: Chapman and Johnson (2002), Keysar and Barr (2002), Epley and Gilovich (2001, 2006).

The *availability* heuristic suggests that individuals place greater emphasis on recent and especially striking events when it comes to assessing the frequency or probability of these events occurring in the future (Tversky & Kahneman, 1974; Camerer & Loewenstein, 2004). For example, an individual might assess the probability of breast cancer among middle-aged women by recalling such occurrences among their friends and acquaintances. Similarly, an agent might evaluate the likelihood that a given business deal might fall through by recollecting various obstacles it might face (Tversky & Kahneman, 1974). The availability heuristic is based on the stock of information an actor already has in store, or has recently acquired, and is, thus, able to remember. Illusory correlation, effectiveness of a search set, and ease of recall and retrievability are some of the biases discussed in relation to the availability heuristic (Tversky & Kahneman, 1974; Kahneman & Tversky, 1996).

The *representativeness* heuristic is another cognitive shortcut that plays an important role in judgments about the likelihood of events. *Representativeness* is defined as "an assessment of the degree of correspondence between a sample population, an instance and a category, an act and an actor or, more generally, between an outcome and a model. The model may refer to a person, a coin, or the world economy, and the respective outcomes could be marital status, a sequence of heads and tails, or the current price of gold" (Tversky & Kahneman, 2002, 22). For example, when evaluating the probability that a given entertainer (musician, actor, etc.) uses illicit drugs, one might consider the similarity between that actor and the stereotypical drug user; in this instance, "the representativeness heuristic piggy-backs on automatic pattern-matching processes" (Gilovich & Griffin, 2002, 3). Similarly, when estimating the popular appeal of a new technological gadget or product, one might reflect on the parallels between that new gadget and successful technologies on the market.

With the *anchoring-and-adjustment* heuristic, actors latch onto or "anchor" on a piece of information that comes easily to mind and then adjust away from that piece of information until they arrive at an acceptable decision or value (Tversky & Kahneman, 1974; Epley & Gilovich, 2006). To illustrate this heuristic, Epley and Gilovich (2006, 311) describe the following scenario: "participants are asked to first make a comparative assessment (e.g., 'Is the population of Chicago more or less than 200,000?') and then to provide an absolute estimate (e.g., 'What is the actual population of Chicago?')." A substantial body of empirical research based on this paradigm demonstrates that agents' final estimates are biased towards the initial "anchor" value (Epley & Gilovich, 2006). This appears to be true whether the "anchor" is self-generated or provided externally, during an experiment, for example. The anchoring and adjustment heuristic has been associated with a wide range of phenomena and biases, including preference reversals, the hindsight bias, and egocentric biases, to name a few (Chapman & Johnson, 2002; Epley & Gilovich, 2006). Although this heuristic can often be useful, research finds that the adjustments actors make are frequently insufficient, resulting in final assessments that are biased in favour of the initial anchor value or piece of information (Epley & Gilovich, 2001, 2006).

It is important to note that heuristics can have both positive and negative effects. As Camerer and Loewenstein (2004, 11) observe, "[a] good heuristic provides fast, close to optimal, answers when time or cognitive capabilities are limited, but it also violates logical principles and leads to errors in some situations". Examining the use of heuristics in inter-firm interactions at trade fairs complements the contributions of the garbage-can model and is helpful in developing meaningful interpretations of search in the "organized anarchy" setting of the trade fair. With respect to the availability heuristic, for example, it is not difficult to imagine how a recent product failure, for instance, might lead economic agents to inflate the probability of a future product failure, thereby influencing the way exhibiting firms interact with customers and competitors during a trade fair. Further, in relation to the availability heuristic, the immediacy of information flows and the ease with which information is retrieved at trade fairs could also strongly shape the nature of the search processes of firms with respect to the "solutions" they come across during these events. The representativeness heuristic can also be envisaged in the context of inter-firm interactions at trade fairs – for example, when firms are assessing the suitability of a potential business partner, or how to package and present their products to appeal more broadly to customers.

4.4.3 Preferences

In addition to the above insights regarding the bounded rationality of actors and the use of heuristic shortcuts in decision-making, empirical research in behavioral economics also challenges standard economic assumptions regarding the nature of individual preferences, preference formation, and change (Kuran, 1997; George, 2001). A popular example of "non-standard" preferences, or preferences that violate standard economic principles, is *other-regarding preferences*. Briefly stated, other-regarding preferences suggest that issues of fairness and the welfare of others play a role in individual preferences; self-interest is not the only factor in the development of individual preferences (Kahneman et al., 1986; Andreoni & Miller, 2002). Similar to the garbage-can model's notion of "problematic preferences", the work of behavioural economists acknowledges that individual preferences sometimes conflict with one another and that the priorities of individuals are also malleable in the sense that they may be continuously redefined and reordered (George, 2001).

When it comes to preferences in decision-making, the behavioural economics literature distinguishes between two different modes of evaluation: joint evaluation (JE) and separate evaluation (SE). The JE mode refers to a situation in which a variety of alternatives are displayed simultaneously and can, therefore, be more readily compared and contrasted; the SE mode, on the other hand, refers to situations in which different options are presented one at a time and, thus, must be assessed in isolation (Hsee, 1996; Hsee et al., 1999). While studies in this area have offered detailed descriptions of, for example, the nature of "preference reversals" in the different evaluation modes (see, for example, Hsee, 1996; Hsee et al., 1999; Lichtenstein

& Slovic, 2006), for present purposes, the critical insight is that the two modes – joint and separate – play an important role in shaping the nature of individual preferences. In other words, the evaluation context itself can significantly influence preference formation and change. While it is beyond the scope of this research to comparatively examine the nature of firms' preferences in joint (i.e., within the trade fair) versus separate (i.e., outside the trade fair) evaluation contexts, the JE-SE distinction is important to keep in mind when exploring the search and information processes of firms at trade fairs – a context in which multiple options are simultaneously presented to firms.

While the above discussion offers a very condensed treatment of recent contributions from the behavioural economics literature, it nevertheless draws attention to some important ideas and insights that are relevant to the present study. The emphasis placed on "bounded rationality", for instance, has important implications for understanding the nature and dynamics of inter-firm interactions at trade fairs. When faced with the uncertainty and complexity of rapidly changing markets, it is to be expected that firms' search processes will involve the use of heuristic devices (e.g., analogical reasoning) to determine what might be useful or important to them. Recognition of the inconsistency and malleability of individual preferences, in general, as well as the interplay between such preferences and different decision-making settings (e.g., joint versus separate evaluation modes) also contributes to the present study's empirical analysis – for example, by supporting and further extending the garbage-can model's notion of problematic preferences.

The application of behavioural economics to organizations is still very much in its infancy (Camerer & Malmendier, 2007; Armstrong & Huck, 2010). Yet, recent steps in this direction draw attention to several important areas of empirical relevance. Although research in this area generally focuses on the non-standard behaviour of consumers and assumes that firms are fully rational, profit-maximizing agents, recent studies demonstrate that there are occasions where firms may also engage in non-standard behaviour (Armstrong & Huck, 2010). Examples include the use of simple "rules-of-thumb", such as imitating the strategies of more successful firms, rather than deliberating over complex optimal strategies; and, following Simon's (1956) satisficing approach to decision-making, there may be times when firms are fairly content with "satisfactory" as opposed to maximum profits (Armstrong & Huck 2010). While this study does not attempt a direct application of behavioural economics, it does allow the ideas from this subfield to further elucidate the nature of firms' search and information processes at trade fairs.

Beyond this, applications of the ideas of behavioural economics to other contexts – such as, securities regulation and, specifically, the Securities and Exchange Commission – suggest the importance of institutions in, at least partially, overcoming cognitive biases and guiding individual decision-making (Choi & Pritchard, 2003). In the present context, I argue that the specific institutional arrangements of the firm's home country may play such a role and, in doing so, encourage firms to follow development pathways that are largely consistent with that country's previously established patterns of technological specialization.

Finally, the methods and approaches employed in the subfield of behavioural economics are deliberately interdisciplinary in nature. Indeed, behavioural economists have been characterized as "methodological eclectics" (Angner & Loewenstein, 2012, 668) that are comfortable employing a variety of methods and different types of evidence. According to Angner and Loewenstein (2012, 674), "the field [of behavioural economics] has been successful in part because it has drawn inspiration from theoretical and methodological developments in neighboring fields, and succeeded in integrating methods from these fields and in exploiting data gathered from a variety of sources". In perhaps more subtle ways, this book also embraces such interdisciplinarity and suggests that there is much to gain from encouraging greater dialogue among the often disconnected fields (or subfields) of inquiry.

4.5 Conclusion

With a view to developing more dynamic approaches to the study of capitalist variety, scholars (Jackson & Deeg, 2006, 2008; Deeg & Jackson, 2007) have stressed the need to incorporate micro, meso, and macro-levels into the analysis. This chapter represents the micro foundations of such a multi-level framework, whereas the meso- and macro-levels have already been discussed in Chaps. 2 and 3, respectively. Drawing valuable insights from three separate, yet complementary literatures – the garbage-can model of organizational choice, research on organizational or technological search, as well as relevant work in behavioural economics – this chapter offers a useful framework for conceptualizing and exploring the search and interaction processes of firms during international trade fairs.

The discussion in this chapter suggests that the garbage-can model is particularly well-suited to capturing the dynamics of inter-firm interaction at trade fairs. For the reasons discussed above, trade fairs can be viewed as an organizational form that reflects the basic properties of an "organized anarchy" – namely, problematic preferences, unclear technologies and processes, and fluid participation. Similar to other complex organizational forms to which the model has previously been applied, the dynamics of firms' search and interactions at trade fairs closely resemble garbage-can decision-making processes. For example, agents at the trade fair engage in search often without a clear or consistent idea of what they are looking for or how best to achieve it (problematic preferences); operating under conditions of considerable complexity and uncertainty, trade fair exhibitors come in with limited knowledge of what other actors at the event are doing and rarely have a clear sense of how best to choose a new technology (unclear technologies and processes); the exhibitors at the trade fair are incredibly diverse and bring with them varying amounts of time, energy, and interest (fluid participation).

In addition to the garbage-can model, this chapter highlights important learnings from the literature on organizational and technological search. The two dimensions of search involving *where* (exploitative-explorative) and *how* (experiential-cognitive) to search are particularly useful in supporting the core ideas of the garbage-can model and further illuminating the nature of firms' search and information processes at trade fairs. Although trade fairs have not been systematically explored in the search literature, extending the focus to these events can also enrich the study of search processes itself. For example, a focus on trade fairs seems particularly helpful in identifying the specific mechanisms by which exploitative and explorative searches may be bridged, perhaps in relation to discussions around the role of innovation intermediaries (Lopez-Vega et al., 2016).

Finally, this chapter identifies a number of key ideas from the behavioural economics literature that help inform the qualitative analysis that will be presented in Chap. 5. These include the notion of bounded rationality, the use of heuristic shortcuts under conditions of bounded rationality, and the nature of preferences, preference formation, and change in such settings. These ideas complement key insights from the garbage-can model and the search literature to develop a useful framework for interpreting the complex micro-level processes of search and inter-firm interaction at trade fairs.

References

Andreoni, J., & Miller, J. (2002). Giving according to GARP: An experimental test of the consistency of preferences for altruism. *Econometrica, 70*(2), 737–753.

Angner, E., & Loewenstein, G. (2012). Behavioral economics. In U. Mäki (Ed.), *Handbook of philosophy of science: Philosophy of economics* (pp. 641–690). Elsevier.

Argote, L., & Greve, H. R. (2007). A behavioral theory of the firm – 40 years and counting: Introduction and impact. *Organization Science, 18*(3), 337–545.

Argote, L., & Miron-Spektor, E. (2011). Organizational learning: From experience to knowledge. *Organization Science, 22*(5), 1123–1137.

Armstrong, M., & Huck, S. (2010). *Behavioral economics as applied to firms: A primer*. CESifo Working Paper Series No. 2973.

Bar-Hillel, M., & Neter, E. (2002). How alike is it versus how likely is it? A disjunction fallacy. In T. Gilovich, D. W. Griffin, & D. Kahneman (Eds.), *Heuristics and biases: The psychology of intuitive judgment* (pp. 82–97). Cambridge University Press.

Barnett, W. P., & Hansen, M. T. (1996). The red queen in organizational evolution. *Strategic Management Journal, 17*(1), 139–157.

Bathelt, H. (2005). Geographies of production: Growth regimes in spatial perspectives 2-knowledge creation and growth in clusters. *Progress in Human Geography, 29*(2), 204–216.

Bathelt, H., & Gibson, R. (2015). Learning in "organized anarchies": The nature of technological search processes at trade fairs. *Regional Studies, 49*(6), 985–1002.

Bathelt, H., & Glückler, J. (2011). *The relational economy: Geographies of knowing and learning*. Oxford University Press.

Bathelt, H., & Schuldt, N. (2008). Between luminaires and meat grinders: International trade fairs as temporary clusters. *Regional Studies, 42*(6), 853–868.

Bathelt, H., Malmberg, H., & Maskell, P. (2004). Clusters and knowledge: Local buzz, global pipelines and the process of knowledge creation. *Progress in Human Geography, 28*, 31–56.

Bathelt, H., Golfetto, F., & Rinallo, D. (2014). *Trade shows in the globalizing knowledge economy*. Oxford University Press.

Benner, M. J., & Tushman, M. (2002). Process management and technological innovation: A longitudinal study of the paint and photography industries. *Administrative Science Quarterly, 47*, 676–698.

Bromiley, P. (1986). Planning systems in large organizations: A garbage can approach with application to defense PPBS. In J. G. March & R. W. Baylon (Eds.), *Ambiguity and command: Organizational perspectives on military decision making* (pp. 120–139). Prentice Hall Press.

Camerer, C. F., & Loewenstein, G. (2004). Behavioral economics: Past, present, future. In C. F. Camerer, G. Loewenstein, & M. Rabin (Eds.), *Advances in behavioral economics* (pp. 1–52). Princeton University Press.

Camerer, C. F., & Malmendier, U. (2007). Behavioral economics of organizations. In P. Diamond & H. Vartiainen (Eds.), *Behavioral economics and its applications* (pp. 235–290). Princeton University Press.

Camerer, C. F., Issacharoff, S., Loewenstein, G., O'donoghue, T., & Rabin, M. (2003). Regulation for conservatives: Behavioral economics and the case for "asymmetric paternalism". *University of Pennsylvania Law Review, 151*(3), 1211–1254.

Camerer, C. F., Loewenstein, G., & Prelec, D. (2004). Neuroeconomics: Why economics needs brains. *The Scandinavian Journal of Economics, 106*(3), 555–579.

Cantwell, J. (1989). *Technological innovation and multinational corporations*. Blackwell.

Chapman, G. B., & Johnson, E. J. (2002). Incorporating the irrelevant: Anchors in judgments of belief and value. In T. Gilovich, D. Griffin, & D. Kahneman (Eds.), *Heuristics and biases: The psychology of intuitive judgment* (pp. 120–138). Cambridge University Press.

Choi, S. J., & Pritchard, A. C. (2003). Behavioral economics and the SEC. *Stanford Law Review, 56*(1), 1–73.

Clark, D. L., et al. (Eds.). (1980). *New perspectives on planning in educational organizations*. Far West Laboratory.

Cohen, M. D., March, J. G., & Olsen, J. P. (1972). A garbage can model of organizational choice. *Administrative Science Quarterly, 17*(1), 1–25.

Cyert, R. M., & March, J. G. (1963). *A behavioral theory of the firm*. Prentice-Hall.

Dalziel, M. (2010). Why do innovation intermediaries exist? In *Proceedings of DRUID summer conference*.

Deeg, R., & Jackson, G. (2007). Towards a more dynamic theory of capitalist variety. *Socio-Economic Review, 5*, 149–179.

Diamond, P., & Vartiainen, H. (Eds.). (2007). *Behavioral economics and its applications*. Princeton University Press.

DiMaggio, P. J., & Powell, W. W. (1983). The iron cage revisited: Collective rationality and institutional isomorphism in organizational fields. *American Sociological Review, 48*(2), 147–160.

Eggink, M. (2013). The components of an innovation system: A conceptual innovation system framework. *Journal of Innovation and Business Best Practices, 2013*, 1–12.

Epley, N., & Gilovich, T. (2001). Putting adjustment back in the anchoring and adjustment heuristic: Differential processing of self-generated and experimenter-provided anchors. *Psychological Science, 12*(5), 391–396.

Epley, N., & Gilovich, T. (2006). The anchoring-and-adjustment heuristic: Why the adjustments are insufficient. *Psychological Science, 17*(4), 311–318.

Fardal, H., & Sørnes, J. O. (2008). IS strategic decision-making: A garbage can view. *Issues in Informing Science & Information Technology, 5*, 553–569.

Fleming, L., & Sorenson, O. (2001). Technology as a complex adaptive system: Evidence from patent data. *Research Policy, 30*(7), 1019–1039.

Fleming, L., & Sorenson, O. (2004). Science as a map in technological search. *Strategic Management Journal, 25*(8–9), 909–928.

Gavetti, G. (2012). PERSPECTIVE – Toward a behavioral theory of strategy. *Organization Science, 23*(1), 267–285.

Gavetti, G., & Levinthal, D. A. (2000). Looking forward and looking backward: Cognitive and experimental search. *Administrative Science Quarterly, 45*(1), 113–137.

Gavetti, G., & Rivkin, J. W. (2007). On the origin of strategy: Action and cognition over time. *Organization Science, 18*(3), 420–439.

Gavetti, G., Greve, H. R., Levinthal, D. A., & Ocasio, W. (2012). The behavioral theory of the firm: Assessment and prospects. *Academy of Management Annals, 6*(1), 1–40.

George, D. (2001). *Preference pollution: How markets create the desires we dislike*. University of Michigan Press.

Gertler, M. S. (1995). 'Being there': Proximity, organization, and culture in the development and adoption of advanced manufacturing technologies. *Economic Geography, 71*(1), 1–26.

Gertler, M. S. (2003). Tacit knowledge and the economic geography of context, or the undefinable tacitness of being (there). *Journal of Economic Geography, 3*(1), 75–99.

Gigerenzer, G., & Todd, P. M. (1999). *Simple heuristics that make us smart*. Oxford University Press.

Gilovich, T., & Griffin, D. (2002). Introduction – Heuristics and biases: Then and now. In T. Gilovich, D. Griffin, & D. Kahneman (Eds.), *Heuristics and biases: The psychology of intuitive judgment* (pp. 1–18). Cambridge University Press.

Helfat, C. E. (1994). Evolutionary trajectories in petroleum firm R&D. *Management Science, 40*, 1720–1747.

Ho, T. H., Lim, N., & Camerer, C. F. (2006). Modeling the psychology of consumer and firm behavior with behavioral economics. *Journal of Marketing Research, 43*(3), 307–331.

Honda, E. H. (2007). *Blame it on Rio: Recycling the paradox of the Kyoto Protocol through the garbage can model*. Annual Meeting of the International Studies Association.

Howells, J. (2006). Intermediation and the role of intermediaries in innovation. *Research Policy, 35*(5), 715–728.

Hsee, C. K. (1996). The evaluability hypothesis: An explanation for preference reversals between joint and separate evaluations alternatives. *Organizational Behavior and Human Decision Processes, 67*(3), 247–257.

Hsee, C. K., Blount, S., Loewenstein, G. F., & Bazerman, M. H. (1999). Preference reversals between joint and separate evaluations of options: A review and theoretical analysis. *Psychological Bulletin, 125*(5), 576–590.

Hsieh, C., Nickerson, J. A., & Zenger, T. R. (2007). Opportunity discovery, problem-solving and a theory of the entrepreneurial firm. *Journal of Management Studies, 44*(7), 1255–1277.

Huber, G. P. (1991). Organizational learning: The contributing processes and the literatures. *Organization Science, 2*(1), 88–115.

Jackson, G., & Deeg, R. (2006). *How many varieties of capitalism? Comparing the comparative institutional analyses of capitalist diversity*. MPIfG Discussion Paper 06/2, Max Planck Institute for the Study of Societies.

Jackson, G., & Deeg, R. (2008). Comparing capitalisms: Understanding institutional diversity and its implications for international business. *Journal of International Business Studies, 39*(4), 540–561.

Kahneman, D., & Frederick, S. (2002). Representativeness revisited: Attribute substitution in intuitive judgment. In T. Gilovich, D. Griffin, & D. Kahneman (Eds.), *Heuristics and biases: The psychology of intuitive judgment* (pp. 49–81). Cambridge University Press.

Kahneman, D., & Tversky, A. (1996). On the reality of cognitive illusions. *Psychological Review, 103*(3), 582–591.

Kahneman, D., Knetsch, J. L., & Thaler, R. (1986). Fairness as a constraint on profit seeking: Entitlements in the market. *The American Economic Review, 76*(4), 728–741.

Katapodi, M. C., Facione, N. C., Humphreys, J. C., & Dodd, M. J. (2005). Perceived breast cancer risk: Heuristic reasoning and search for dominance structure. *Social Science and Medicine, 60*(2), 421–432.

Katila, R., & Ahuja, G. (2002). Something old, something new: A longitudinal study of search behavior and new product introduction. *Academy of Management Journal, 45*(8), 1183–1194.

Keysar, B., & Barr, D. J. (2002). Self-anchoring in conversation: Why language users do not do what they 'should'. In T. Gilovich, D. W. Griffin, & D. Kahneman (Eds.), *Heuristics and biases: The psychology of intuitive judgment* (pp. 150–166). Cambridge University Press.

Kingdon, J. W. (2003). *Agendas, alternatives and public policies* (2nd ed.). Addison-Wesley Educational Publishers.

Kuran, T. (1997). *Private truths, public lies: The social consequences of preference falsification.* Harvard University Press.

Laursen, K. (2010). Do export and technological specialisation patterns co-evolve in terms of convergence or divergence? Evidence from 19 OECD countries, 1971-1991. *Journal of Evolutionary Economics, 10*, 415–436.

Laursen, K. (2012). Keep searching and you'll find: What do we know about variety creation through firms' search activities for innovation? *Industrial and Corporate Change, 21*(5), 1181–1220.

Laursen, K., Leone, M. I., & Torrisi, S. (2010). Technological exploration through licensing: New insights from the licensee's point of view. *Industrial and Corporate Change, 19*(3), 871–897.

Levinthal, D., & March, J. G. (1993). The myopia of learning. *Strategic Management Journal, 14*, 95–112.

Levinthal, D., & Rerup, C. (2006). Crossing an apparent chasm: Bridging mindful and less-mindful perspectives on organizational learning. *Organization Science, 17*(4), 502–513.

Levitt, B., & March, J. G. (1988). Organizational learning. *Annual Review of Sociology, 14*(1), 319–338.

Lichtenstein, S., & Slovic, P. (Eds.). (2006). *The construction of preference*. Cambridge University Press.

Lindblom, C. (1959). The science of 'muddling through'. *Public Administration Review, 19*(2), 79–88.

Lomi, A., & Harrison, J. R. (2012). The garbage can model of organizational choice: Looking forward at forty. In A. Lomi & J. R. Harrison (Eds.), *The garbage can model of organizational choice: Looking forward at forty* (pp. 3–17). Emerald Group Publishing Limited.

Lopez-Vega, H., Tell, F., & Vanhaverbeke, W. (2016). Where and how to search? Search paths in open innovation. *Research Policy, 45*(1), 125–136.

March, J. G. (1991). Exploration and exploitation in organization learning. *Organization Science, 2*(1), 71–87.

March, J. G., & Olsen, J. P. (1984). The new institutionalism: Organizational factors in political life. *The American Political Science Review, 78*(3), 734–749.

March, J. G., & Simon, H. A. (1958). *Organizations*. Wiley.

March, J. G., & Weissinger-Baylon, R. (Eds.). (1986). *Ambiguity and command: Organizational perspectives on military decision making*. Pitman.

Martin, X., & Mitchell, W. (1998). The influence of local search and performance heuristics on new design introduction in a new product market. *Research Policy, 26*, 753–771.

Maskell, P., Bathelt, H., & Malmberg, A. (2006). Building global knowledge pipelines: The role of temporary clusters. *European Planning Studies, 14*, 997–1013.

Meyer, J. W., & Rowan, B. (1977). Institutionalized organizations: Formal structure as myth and ceremony. *American Journal of Sociology, 83*(2), 340–363.

Miner, A. S., Amburgey, T. L., & Stearns, T. M. (1990). Interorganizational linkages and population dynamics: Buffering and transformational shields. *Administration Science Quarterly, 35*(4), 689–713.

Nelson, R. R., & Winter, S. G. (1982). *An evolutionary theory of economic change*. Harvard University Press.

Nerkar, A., & Roberts, P. W. (2004). Technological and product-market experience and the success of new product introductions in the pharmaceutical industry. *Strategic Management Journal, 25*(8–9), 779–799.

Newmann, W. W. (1998). Foreign policy decision making, garbage cans, and policy shifts: The Eisenhower administration and the 'chances for peace' speech. *American Review of Public Administration, 28*, 187–212.

Nickerson, J. A., & Zenger, T. R. (2004). A knowledge-based theory of the firm: The problem-solving perspective. *Organization Science, 15*(6), 617–632.

Pavitt, K. (1988). International patterns of technological accumulation. In N. Hood & J.-E. Vahlne (Eds.), *Strategies in global competition* (pp. 126–157). Croom Helm.

Phillimore, J., & Joseph, R. (2003). Science parks: A triumph of hype over experience? In L. V. Shavinina (Ed.), *The international handbook on innovation* (pp. 750–757). Elsevier Ltd.

Pisano, G. P. (1994). Knowledge, integration, and the locus of learning: An empirical analysis of process development. *Strategic Management Journal, 15*(S1), 85–100.

Postrel, S. (2002). Islands of shared knowledge: Specialization and mutual understanding in problem-solving teams. *Organization Science, 13*(3), 303–320.

Radner, R. (1996). Bounded rationality, indeterminacy, and the theory of the firm. *The Economic Journal, 106*(438), 1360–1373.

Rosenkopf, L., & Nerkar, A. (2001). Beyond local research: Boundary-spanning, exploration, and impact in the optical disk industry. *Strategic Management Journal, 22*(4), 287–306.

Schuldt, N., & Bathelt, H. (2011). International trade fairs and global buzz. Part II: Practices of global buzz. *European Planning Studies, 19*(1), 1–22.

Schwarz, N., & Vaughn, L. A. (2002). The availability heuristic revisited: Recalled content and ease of recall as information. In T. Gilovich, D. W. Griffin, & D. Kahneman (Eds.), *Heuristics and biases: The psychology of intuitive judgment* (pp. 103–119). Cambridge University Press.

Simon, H. A. (1955). A behavioral model of rational choice. *The Quarterly Journal of Economics, 69*(1), 99–118.

Simon, H. A. (1956). Rational choice and the structure of the environment. *Psychological Review, 63*(2), 129.

Simon, H. A. (1987a). Behavioral economics. In J. Eatwell, M. Milgate, & P. Newman (Eds.), *The new Palgrave: A dictionary of economics* (Vol. 1, pp. 221–225). Macmillan.

Simon, H. A. (1987b). Satisficing. In J. Eatwell, M. Milgate, & P. Newman (Eds.), *The new Palgrave: A dictionary of economics* (Vol. 1, pp. 243–245). Macmillan.

Spradlin, D. (2012). Are you solving the right problem? *Harvard Business Review, 90*, 84–93.

Starmer, C. (2004). Developments in non-expected utility theory: The hunt for a descriptive theory of choice under risk. In C. F. Camerer, G. Loewenstein, & M. Rabin (Eds.), *Advances in behavioral economics* (pp. 104–147). Princeton University Press.

Stuart, T., & Podolny, J. (1996). Local search and the evolution of technological capabilities. *Strategic Management Journal, 17*, 21–38.

Tripsas, M., & Gavetti, G. (2000). Capabilities, cognition, and inertia: Evidence from digital imaging. *Strategic Management Journal, 21*(10/11), 1147–1161.

Tversky, A., & Kahneman, D. (1974). Judgment under uncertainty: Heuristics and biases. *Science, 185*(4157), 1124–1131.

Tversky, A., & Kahneman, D. (1983). Extensional versus intuitive reasoning: The conjunction fallacy in probability judgment. *Psychological Review, 90*(4), 293.

Tversky, A., & Kahneman, D. (2002). Extensional versus intuitive reasoning: The conjunction fallacy in probability judgment. In T. Gilovich, D. W. Griffin, & D. Kahneman (Eds.), *Heuristics and biases: The psychology of intuitive judgment* (pp. 19–48). Cambridge University Press.

Weiner, T. (2007). Touching the third rail: Explaining the failure of Bush's social security initiative. *Politics and Policy, 35*(4), 872–897.

Williamson, O. E. (1985). *The economic institutions of capitalism*. Free Press.

Wilson, J., & Joseph, J. (2015). Organizational attention and technological search in the multibusiness firm: Motorola from 1974 to 1991. In W. Ocasio (Ed.), *Cognition and strategy (Advances in strategic management, Volume 32)* (pp. 407–435). Emerald Group Publishing Limited.

Yang, H., Steensma, H. K., & Phelps, C. (2010). Learning from what others have learned from you: The effects of knowledge spillovers on originating firms. *Academy of Management Journal, 53*(2), 371–389.

Chapter 5
The Nature of Firms' Search Processes at International Trade Fairs: A Qualitative Analysis

5.1 Introduction

This chapter represents the first part of the study's empirical analysis. Adopting a qualitative approach, its main objective is to elucidate the complex and multi-faceted processes of search that are underway during leading trade fairs. To this end, it focuses specifically on the processes by which firms identify and ultimately choose solutions – be they in the form of a new technology, a new product or process innovation, or a potential business partner. Of particular interest are the motivations underlying a firm's search for these solutions, the manner in which it goes about searching for them, and whether such searches occur in knowledge terrains that are largely familiar or outside of the firm's existing knowledge bases. The chapter demonstrates that: (i) the search processes of firms at trade fairs are rarely oriented by a clearly defined goal or set of objectives; (ii) the search processes of firms at these events tend to be shaped by the particular context in which the firm operates; and (iii) the searches taking place at trade fairs generally lead to the adoption of technologies that are actually very close to the firm's existing technologies and production context.

To guide the interpretation of the qualitative analysis, I rely, first and foremost, on the concept of "organized anarchies" and the garbage-can model of organizational choice (Cohen et al., 1972), which was discussed in Chap. 4. The model's emphasis on the non-linear, often ambiguous dynamics of decision-making in organized anarchies, along with the multiple process streams – problems, solutions, participants – that flow through these complex organizational settings are especially salient to the forthcoming analysis. Yet, beyond the garbage-can model, this chapter also draws on the technological search literature and key ideas from the subfield of behavioural economics, which were also presented in Chap. 4. Once the micro processes of search have been identified and explained in this chapter, the central task remaining will be to generalize these findings to a broader set of firms from multiple

R. Gibson, *International Trade Fairs and Inter-Firm Knowledge Flows*,
https://doi.org/10.1007/978-3-031-20557-6_5

trade fairs, and then link these micro-level processes to ongoing discussions regarding the persistence or decline of distinct capitalist varieties. This latter task will be the focus of Chap. 6.

Based on the garbage-can model (Cohen et al., 1972), this chapter hypothesizes that the search behaviours of firms at trade fairs will rarely follow a linear or rational trajectory; instead, these searches will more closely resemble garbage-can-like processes – that is, processes in which "problems, solutions, and participants move from one choice opportunity to another in such a way that the nature of the choice, the time it takes, and the problem it solves all depend on a relatively complicated intermeshing of elements" (Cohen et al., 1972, 16). Moreover, within this complex organizational setting, it is expected that firms' search behaviours will be conditioned by their responses to the constraints imposed by their own bounded rationality (Simon, 1955, 1987a). This is expected to play out in different ways, with some firms relying on different heuristic short-cuts, for instance. Yet, the overarching hypothesis is that the search processes of firms will invariably be shaped by their pre-existing organizational routines, production contexts, and knowledge bases, all of which are associated with the institutional make-up of the country in which these firms operate. From this, it seems reasonable to speculate that the search and interaction processes of firms at trade fairs will generally support continued divergence in the technological specializations of national states, rather than setting them on a pathway towards increasing convergence on a single national model.

The empirical setting for the forthcoming qualitative analysis consists of four leading North American trade fairs – two in the U.S. and two in Canada. As previously described, a total of 65 semi-structured interviews specifically focused on search processes were conducted with exhibitors at these four events. The findings from these interviews form the basis of an original typology that classifies different types of search in relation to: (i) the goal underlying the search process and (ii) the specific approach or strategy used to search. Through the development and application of this typology, the qualitative study contributes new empirical evidence on the micro-level processes of knowledge exchange at trade fairs. In so doing, a key aim is to create closer linkages between the micro processes of firm-level search at trade fairs and broader macro-level shifts that may be taking place in areas such as national production, technological focus, and innovation patterns. Establishing such linkages can offer important clues as to why firms may choose different pathways in technological production, how these choices could potentially lead to incremental forms of institutional change, and yet why the selected pathways observed in this study seem to reinforce rather than undermine existing national specializations.

The remainder of the chapter proceeds as follows. The next section introduces and describes a typology of the search processes of firms at trade fairs. This section includes an overview of the steps taken to develop the typology, a brief discussion of the prevalence of search at trade fairs, and a comprehensive description of the typology, which includes illustrative examples from the interview transcripts. The following section focuses on the results of the typology before turning, in the next section, to a discussion of the key findings in relation to the study's overarching research objectives. The chapter concludes by summarizing the results of the qualitative analysis and discussing how the quantitative portion of the analysis, presented in the next chapter, will extend and provide a broader empirical foundation for these findings.

5.2 A Typology of Search Processes at Trade Fairs[1]

The typology of search processes presented in this study identifies four types of firm-level searches: *Type 1 – Unclear and Unstructured*; *Type 2 – Ambiguous but Optimistic*; *Type 3 – Confident yet Unclear*; and *Type 4 – Clear and Structured.* The typology was constructed through a multi-stage approach: To begin, I went through each of the interviews in a general way, identifying and making notes on common themes, frequently mentioned words, common expressions, and so on. With this broad guideline in mind, I then proceeded to examine the individual interview transcripts in greater detail, placing each in an initial category according to its stated search goals and search methods/strategies. At this stage, tentative names for the different search types were assigned – for example, "unclear goals" and "unclear methods". The next careful reading of the transcripts helped further clarify the number and contours of the categories as well as the location of each interview within these constructed search types. An outside reader was then assigned to conduct a "blind" review of the interview transcripts to check whether there were any discrepancies in their initial assignment to the specific categories.

5.2.1 *The Prevalence of Search Processes at Trade Fairs*

Before looking more closely at the nature of the different search categories and their relationship to other dimensions, it is important to first clarify that most exhibiting firms do not participate in trade fairs solely for the purpose of displaying their own products or generating sales or leads. As Table 5.1 indicates, the majority of trade fair exhibitors, which are the focus of the interviews, are also interested in examining the exhibits of other firms and informing themselves about new and emerging trends, markets, and technologies. As one respondent from a firm in the metalworking industry explained, "[w]e try to see who has got something that is faster, better, quicker. And if something catches our eye that is similar to the industry that we are in or that we are targeting, then, for sure, we will stop by and ask some questions." This statement supports the idea of the dual role of trade fair exhibitors. Rather than

[1] This section draws from and expands the findings presented in Bathelt, H. and Gibson, R. (2015). Learning in 'organized anarchies': The nature of technological search processes at trade fairs, *Regional Studies* 49(6), 985–1002. Copyright Regional Studies Association. Reprinted by permission of Taylor & Francis Ltd., http://www.tandfonline.com, on behalf of Regional Studies Association.

Table 5.1 Characteristics of search processes at Four North American Trade Fairs

Trade fair	Exhibiting firms that informed themselves about other exhibits at the trade fair % (n)	Exhibiting firms that spent at least 1 h per day inspecting other exhibits % (n)	Exhibiting firms that did not have a particular goal when searching other exhibits % (n)
LFI, New York	94% (16)	76% (13)	35% (6)
NPE, Chicago	100% (19)	64% (12)	53% (10)
CMTS, Toronto	100% (17)	50% (8)	53% (9)
MMTS, Montreal	100% (12)	42% (5)	33% (4)
Total	99% (64)	59% (38)	45% (29)

Notes: *LFI* LightFair International, *NPE* International Plastics Showcase, *CMTS* Canadian Manufacturing Technology Show, *MMTS* Montreal Manufacturing Technology Show

simply showcasing their latest products and technologies, exhibitors are also like visitors in the sense that they have a clear focus on learning during the trade fair (Bathelt & Gibson, 2015). In addition to searching for new market or technological trends, many firms also search for new business partners at the trade fair. The following quote from a composites manufacturing firm nicely illustrates this point: "Actually, we try to have at least one guy walking around all the time [...] because some of those booths are our potential partners as well. We want to talk to those guys and make some contact. It's not only about finding customers; it's also finding partners. [...] It is as important as finding customer leads."

Table 5.1 demonstrates that trade fair participants generally spend a considerable amount of time exploring the other exhibit booths at the trade fair. For example, during the U.S. design fair, LightFair International (LFI), firms set aside an average of 2.5 h per day to walk around the exhibit hall examining other booths, with over three-quarters of firms devoting at least 1 h to this activity. At the International Plastics Showcase (NPE), in Chicago, approximately two-thirds of firms said they devoted at least 1 h per day to inspecting other exhibits. As might be expected, firms at the smaller trade fairs, such as the Canadian Manufacturing Technology Show (CMTS) and the Montreal Manufacturing Technology Show (MMTS), typically spent less time examining other exhibits. Based on the interview data, only 50% of CMTS exhibitors and 42% of MMTS exhibitors devoted at least 60 min per day to such search activities. Still, these average estimates indicate a fairly substantial investment of time into search activities during the trade fair given that the exhibition space is smaller and there are fewer exhibitors.

Although trade fairs are widely used to search the other exhibits, the firms engaged in such searches did not always know what they were searching for during the trade fair. As indicated in Table 5.1, a total of almost half (45%) of the firms across all four trade fairs lacked a specific goal when inspecting the other exhibits. Given the high cost of participating in trade fairs and the fact that some events only take place once every 2 or 3 years, it might be expected that more firms would come to the trade fair with a clearer idea of what they hope to achieve through their participation at the event. In marketing and business administration studies, there is quite a substantial body of literature focused on how to get the most out of trade fair participation, including recommendations to exhibitors on how best to prepare themselves for success at such events (see, for example, Gopalakrishna & Williams, 1992; Sashi & Perretty, 1992; Shoham, 1999). Despite this emphasis in the literature, engaging in extensive pre-event preparations and developing clear search strategies were fairly uncommon amongst the 65 firms that were interviewed in the qualitative portion of this study.[2]

5.2.2 *Defining the Typology*

The typology of search processes developed in this study ranges from unclear and unstructured, at the one extreme, to clear and highly structured, at the other end. A description of each of the four different search types follows below, while Table 5.2 provides a detailed overview of the typology and the relationship of the different search types to some basic trade fair characteristics.

Search Type 1: Unclear and Unstructured At the least structured end of the spectrum, firms engage in searches without having clear or pre-defined ideas about what they are actually searching for. The respondents that fell into this category readily admitted this lack of clarity in their search processes and often suggested that they were simply walking around the exhibits, looking to see if something new or interesting might "catch our eye". These firms did not have a clear sense of what this "something" might be or how they might go about finding it. Their goal, for example, was simply "[t]o see some new products in the services. But we are not looking for anything specific." As a result, the firms that fit into this category did not proceed with any kind of pre-established plan. Instead, they would simply "wander around" the exhibition halls to see what might be new or interesting to them. This search type was the most highly represented in the analysis, with close to half (45%) of all firms expressing a lack of clarity in both the goals and structure of their search processes at trade fairs.

As alluded to above, the ambiguity around search goals among *Type 1* firms was generally accompanied by a lack of clarity about how to engage in search activities during the trade fair. This sentiment is nicely captured in the following statement: "Nothing specific. We just try to catch up with people we know and get a sense of what is going on in the industry." Another firm specializing in electronics

---

[2] In interviewing the exhibitors, it became clear that most respondents did not intend to purchase new products or technologies at the actual trade fair. This finding supports the work of Godard and O'Connor (2001), which shows that trade fair visitors are rarely focused on buying products during trade fairs. A study by Borghini et al. (2006) also supports this observation, as it suggests that trade fair attendees are generally more interested in acquiring information than purchasing new products or technologies.

Table 5.2 Typology of firms' search behaviours at North American Trade Fairs

Category	Type 1	Type 2	Type 3	Type 4
	Unclear and unstructured	Ambiguous but optimistic	Confident but unclear	Clear and structured
Description	No clear goal in searching the other exhibits; no evidence of clear approach to searching	Only a very general goal; but optimistic they will recognize what is important or useful when they come across it	Insist they know what they are looking for; but no articulation of the precise nature of the goal or approach to searching	Articulation of a clear goal; indication of a systematic approach to searching
Firm interview ID	CMTS 11, 12, 13, 14, 15, 20, 31, 38, 57	CMTS 21	CMTS 39, 46, 50, 51, 52	CMTS 47, 49
	MMTS 08, 21, 26, 27	MMTS 06	MMTS 07, 17, 20, 32	MMTS 09, 10, 33
	NPE 24, 26, 27, 28, 29, 36, 37, 44, 45, 63		NPE 23, 25, 31, 32, 62, 69, 70	NPE 17, 18
	LFI 15, 23, 27, 35, 43, 48	LFI 18, 19, 31, 53	LFI 14, 20, 21, 34, 44, 47	LFI 13
Regional trade fair (number of firms)	13	2	9	5
International trade fair (number of firms)	16	4	13	3
Design focus (number of firms)	6	4	6	1
Technology focus (number of firms)	23	2	16	7
Total (n = 65)	29	6	22	8
Share of firms	45%	9%	34%	12%

Notes: *LFI* LightFair International, *NPE* International Plastics Showcase, *CMTS* Canadian Manufacturing Technology Show, *MMTS* Montreal Manufacturing Technology Show

components and supplies suggested that there *was* a specific way to search at these events, but that this differed depending on the goal: "There is a specific way to go and see our competitors – what they are doing. There is also another way to see what is new, what is out there, and [what is] supporting our products." While this statement seems to suggest a lack of fit with the *Type 1 (unclear and unstructured)* search category, when probed more deeply, there was actually very little clarity on what these "specific" ways were or what the underlying goals of search might be in the first place. The same firm revealed the following: "No. It's all open. We just want to see what is out there." In other words, there was a general spirit of openness and non-linearity characterizing the search processes of firms that fit into the *Type 1* search category. This overarching quality is nicely captured in the following statement from a thermoplastic injection molding firm: "You don't always know what you are looking for; you might just see something that is new and cool and wonder how you might apply it to your own products." This sentiment strongly resonates with the garbage-can model's description of "organized anarchies", particularly with respect to the core properties of problematic preferences and unclear technology or processes. It also clearly points to the role of the firm's existing production context and knowledge bases in determining what might be useful or important to the firm – specifically the latter part of the statement regarding "how you might apply it to your own products". These arguments will be developed more fully in the discussion section of this chapter.

Search Type 2: Ambiguous Yet Optimistic On the face of it, the firms that fell into the second search category did not seem a great deal clearer than those situated in *Search Type 1 (unclear and unstructured)* with respect to their underlying search objectives. Indeed, many firms in the *Type 2* category admitted that they did not know exactly what they were looking for at the trade fair. This level of uncertainty is well articulated in the following quote from a lighting components firm: "Yes and no. It depends on who is the person that is out there looking. Sometimes we are looking for people to partner up with, sometimes we are looking for potential customers, and sometimes we just want to see what is out there." While there is certainly some ambiguity in the above statement, it is, nevertheless, qualitatively different from those that say: "no, it's all open" or "not really, we just look around." In other words, there is a sense in the *Type 2 (ambiguous but optimistic)* quote presented above that there are contingencies at play (e.g., who is searching, what the needs of the firm are at a specific point in time), which make clarity regarding the goals of search more difficult to achieve or even articulate. These elements of complexity and contingency serve as good examples of two of the characteristic features of organized anarchies – specifically, "fluid participation" and "problematic preferences" (Cohen et al., 1972).

Additionally, when respondents were asked how they might determine the potential value of an artifact or another kind of "solution" at the trade fair, the firms that fit into *Search Type 2 (ambiguous yet optimistic)* displayed a certain element of optimism that they would be able to recognize something important when they came across it. Consider the following statement from an established custom lighting firm: "For me, it is usually something that's novel, something, new, something I've never seen before. So, an innovative way to solve some sort of problem – that's what catches my eye. I'm kind of a bit of a problem-solver so I see something and think, oh, interesting, good idea, kind of thing." This response is substantively different from the *Type 1* search processes identified above. That being said, the *Type 2 (ambiguous but optimistic)* search type was more difficult to place, as such firms seemed to fall into other search categories as well. Consider the following example from a medium-sized lighting firm:

> Personally, I would have several goals: to educate myself as to what is new, the latest thinking in product design and product development, but also in technology. So, I have two goals: to learn what's new from a technological standpoint in sources and control here, and then also what our competitors are doing in terms of product development, product design and optical system design. For me it is more general information. I go through the entire show scanning quickly each aisle and, if something interests me, I would focus in on that. I have been in the industry for a long time. You can see just by looking at a product if it is doing something unique optically, for example, or from an aesthetic standpoint you can tell immediately if it is new thinking or a new design direction [...].

While the above statement starts by suggesting there are multiple goals, these goals are actually quite general in nature – for instance: "For me it is more general information." And while the search methods are more specific and more clearly articulated than *Type 1 (unclear and unstructured)* search processes, they are still quite broad and largely unsystematic in nature – for example: "I go through the entire show scanning quickly each aisle and, if something interests me, I would focus on that." But the key point that distinguishes this second search type from *Type 1* is the presence of optimism. This is exemplified in the latter part of the long quote above: "You can see just by looking at a product if it is doing something unique optically, for example, or from an aesthetic standpoint you can tell immediately if it is new thinking or a new design direction."

Search Type 3: Confident Yet Unclear The firms that fit into this third type of search were defined by a high degree of confidence in terms of having a clear idea of what they were looking for. Yet, at the same time, the responses of these firms actually provided little evidence to support these ideas. *Type 3* firms were quite definite and emphatic in their responses, frequently using expressions that involved the word "absolutely" when asked whether they had a particular goal in mind when searching through the exhibits. Consider the following statement by a firm in the metalworking industry: "Absolutely. We try to see who has got something that is faster, better, quicker." In a similar vein, the regional manager of a micro manufacturing firm suggested the following: "Yeah, absolutely, absolutely. I am a consumer of these products as well as the manufacturer of these products so I am also interested to go around and see what's on the market for me. I see what catches my eye. Those are my competitors: What do they offer? What are the prices of their products? And, maybe I can be a sub-dealer of some products that we do not have enough for the plan – something like that."

Despite the apparent expressions of confidence noted above, when *Type 3 (confident yet unclear)* firms were asked to elaborate on their search goals and the processes by which they aimed to achieve such goals, they were rarely able to articulate a systematic approach regarding the search method. In the following case, it is much like *Type 1 (unclear and unstructured)* in the sense that these firms seemed to be simply looking around to see if something might "catch their eye". As indicated by the following *Type 3* firm focused on metal supplies, the goal was simply to find *anything* new: "Yes. Anything new. Anything that we have not seen before. It is usually just something that catches our eye."

Although the above responses are similar to *Search Type 1 (unclear and unstructured),* the major point of difference lies in the firm's level of confidence. The *Type 1* firms openly admitted to not having a clear idea about what they were looking for, whereas *Type 3* firms insisted that there were clear objectives in their searches, despite demonstrating minimal evidence to support such claims. Yet, in addition to having trouble backing-up their confident statements about specific goals, *Type 3 (confident but unclear)* firms were unable to describe a systematic approach in their method of searching. Consider the following statement from an established lean manufacturing firm:

> Oh yes, absolutely. During the day actually, when we see if there is a bit of a down time, we will maybe pick an aisle and go see other guys. Now the thing is, coming to this thing many times, we know a lot of the players and we've kind of made acquaintances. It's good to see old faces and also when we go talk to them, we can very very quickly find out what new innovations they have come up with.

In the above quote, not only is there a lack of precision regarding the actual methods by which the firm engaged in search, it is unclear how much time or even whether concerted search processes would take place during the event. From the above statement, search seems to be an activity that may happen "if there is a bit of down time". From this, it is not surprising that a clear search plan had not been devised prior to the firm attending the trade fair.

Search Type 4: Clear and Highly Structured The last type of search behaviour reveals both a clear focus and a highly systematic approach to searching during the trade fair. The firms that fell into this category were able to refer to specific search objectives and also demonstrated a clear idea of how they would meet them. For example, *Type 4* firms often developed detailed lists or plans prior to coming to the trade fair. They also consulted with different parts of the firm to adapt existing plans when necessary and convened at the end of each day at the trade fair to discuss individual experiences and progress. Moreover, the firms that demonstrated this type of clear and highly structured search often had a post-event plan for evaluating the outcomes of the trade fair in relation to their pre-defined objectives. The following quote from a medium-sized motion control equipment firm nicely captures the essence of the *Type 4 (clear and structured)* search process: "Yes. We do a lot of homework before we actually meet with people and come to the fair. I know exactly where I'm going – I have taken the map. I have narrowed it down to, say, there might be 750–800 exhibitors, and I've narrowed it down and I'm basically only interested in looking at about 12–14 [exhibit booths]. I do all that homework before the event." In a similar vein, the representative of a milling machines and equipment firm described the search process in the following terms: "Typically I have a plan in mind. When I walk out of the booth, I have a list of places I want to go to. I look ahead of time into planning. [...] in my company, we already have been able to look at different processes in our company that may need improving. And in every trade show there is something there that might be of interest, whether to be part of our machine shop or foundry, some need that we may have that can relate. I am always looking out for that." While clearly distinct from the other search types, there is still a strong sense here that the context of the firm plays an important role in guiding the search process. This is exemplified in the latter part of the above quote: "[…] to be part of our machine shop or foundry, some need that we may have that can relate. I am always looking out for that."

5.3 Technological Search Behaviour at Trade Fairs and the Garbage-Can Model

Having provided a detailed description of the four categories of search observed at the selected trade fairs, Table 5.2 summarizes the results of the typology. The main focus here is on the distribution of firms among the different search categories. But the table also links the different categories of search to the international or regional character of the trade fair, as well as the design or technology orientation of the event. Of the 65 firms interviewed across the four North American trade fairs, 29 (45%) were placed into *Type 1 (unclear and unstructured)*, 6 (9%) were placed into *Type 2 (ambiguous but optimistic)*, 22 (34%) were put into *Type 3 (confident but unclear)*, and 8 (12%) were placed into *Type 4 (clear and structured)*. With the exception of *Type 2*, which has the smallest number of firms, all four of the trade fairs were well represented across the search types. In other words, I did not see firms from the Canadian Manufacturing Technology Show (CMTS), for instance, or the LightFair (LFI) cluster in a specific search type. There was a fairly even distribution of trade fairs across the different search types.

Before engaging more deeply with the results of the typology, it is important to briefly discuss some of the choices that were made in allocating firms to the different search categories. With respect to the *goals* of search, for example, the results presented in Table 5.2 suggest that the majority of firms (54%) did not have a clearly articulated goal in mind when scanning the trade fair and inspecting the other exhibits. This number was arrived at by combining the two categories of search located on the unstructured end of the spectrum – *Type 1 (unclear and unstructured)* and *Type 2 (ambiguous but optimistic)*. The firms placed into *Type 1* were generally very candid about not knowing the motives behind their own search behaviours. Most of these firms would simply reply "No" or "no idea" when asked whether they had a particular goal in mind while searching through the trade fair. Alternatively, they sometimes provided an expression along the following lines: "No, it's all open. [We] just want to see what's out there." The relatively few firms that fit into *Type 2 (ambiguous but optimistic)* were less forthright in answering the question about search goals, but similarly vague when it came to identifying the core objectives driving their search activities at trade fairs. That being said, the firms in the *Type 2* category were decidedly more optimistic than were *Type 1* firms, specifically with respect to being able to recognize the potential importance of a solution or artifact when they came across it. Thus, while the two groups are not identical, they both indicate a considerable degree of uncertainty regarding the objectives underlying their search behaviours and can be grouped together on this particular dimension – that is, the goals of search.

The difference between the first two categories and *Type 3 – (confident but unclear)* is also worth clarifying. There is an obvious difference between firms that unequivocally state that they have "no idea" about the goals underlying their search behaviours and those that say: "yes, absolutely". Yet, upon further examination, the difference may not be as stark as it first appears. Despite the tone of self-assurance in the responses, the goals of the firms that were supposedly confident in their objectives were diverse and often poorly articulated. Some *Type 3* firms were preoccupied with what their competitors were doing, whereas the search behaviours of others revolved around looking for new partners or suppliers to support improvements to their existing products or equipment. Still others said that they were simply looking for new technological innovations or solutions.

Confidence clearly defines *Type 3* search processes. Yet, compared to the fourth and most structured category, *(clear and structured),* the firms that fell into *Type 3* lacked precision in describing the goals and methods underlying their search processes. This distinction is well captured in the following statement by the CEO of a small machine manufacturing firm: "Yes, we have objectives on the customer lead side as well as on the partnership side. So, yes we have plans and we have worked with the marketing company to be apprised of the show and to make sure that we have our objectives and make sure that we have those leads and people to follow up afterwards." While the methods of *Type 4* firms are much more clearly defined, the goals still have a tendency to be more implicit than explicit in nature. Thus, in an overarching sense, the underlying motivations of firms' search processes seem more or less in line with the garbage-can model's notion of problematic preferences.

It is important to note that while there may be some overlap in the classifications described above, the implications of the typology remain robust even in the face of some potential shifts in the distribution of firms within each search type. From a theoretical standpoint, for example, the typology of search processes assists in developing a deeper understanding of the phenomenon of trade fairs in the global knowledge economy through the development of clearer conceptualizations and indicators regarding the technological search behaviours of firms. The typology's utility also lies in the clustering of such indicators into categories for exploration and comparison in future empirical work.

5.3.1 General Results of the Qualitative Study

The results displayed in Table 5.2 are striking and suggest that the search processes of firms at trade fairs are rarely motivated by a clearly defined or pre-established set of goals. In the organized anarchy of the trade fair, these processes are pursued by "boundedly rational" economic agents that engage in "satisficing" (Simon, 1955,

1956, 1987b) behaviours in order to arrive at decisions during the event. In the context of this research, international trade fairs have been described by firms as complex, fluid, and often uncertain environments. Consequently, many trade fair participants were somewhat reluctant to come to the fair with a single, predefined goal. According to many, this is deemed too difficult, too limiting, and possibly even counter-productive. The following statement from a large commercial lighting design firm nicely captures the multiple goals underlying the firm's search and information processes at trade fairs: "Personally, I have several goals: to educate myself about what is new, the latest thinking in product design and product development, but also in technology. So, I have multiple goals: to learn what is new from a technological standpoint in sources and control here, and then also what our competitors are doing in terms of product development, product design, and optical system design."

The results of the present analysis resonate with key findings from the literature on technological search – specifically, with the argument that firms generally engage more in exploitative than explorative search processes (Pavitt, 1988; Helfat, 1994; Stuart & Podolny, 1996; Tripsas & Gavetti, 2000; Laursen, 2012). As illustrated in the following quote from a firm specializing in remote source lighting, the search processes of firms at trade fairs were often guided by knowledge from their own production contexts and existing organizational practices: "We know the certain things that we are using in our products, but we are looking for the next generation of those types of products. For example, new power supplies and dimming control models, and such. Really, we have a list of specific technologies that we are looking for." In a similar vein, consider the following response from a small thermoplastics firm: "I am primarily looking for products that relate or could be applied to my company, reduce costs, or bring it up to the next level of technological advancement." The following quote from a software firm in the plastics industry also speaks to the contextualized nature of search at trade fairs: "There is no real difference in tasks for different expertise, but the development guys are more interested, since we are a software company, in software applications or how different machines might be controlled with software." With respect to the context-driven nature of searches, several other firms provided statements along the following lines: "we try to find something that will fit into our strengths" or "what is out there and supporting our products."

The above discussion does not suggest that firms were exclusively engaged in exploitative searches during the trade fair; there was also some evidence of explorative search processes taking place. Several interviewees, for example, stressed the importance of finding something "new" or "novel" through their engagement in search at these events. This sentiment is well captured in the following description by a firm specializing in primary processing equipment for the plastics industry: "I like to go out and see what kinds of new technologies not only my kind of competition is offering, but also new technologies that can help expand and grow our company and make our company more efficient." Another well-established firm focused on custom lighting described the search for new knowledge at trade fairs in the following way:

> For me, it is usually something that's novel, something, new, something I've never seen before. So, an innovative way to solve some sort of problem – that's what catches my eye. I'm kind of a bit of a problem-solver so I see something and think, oh, interesting, good idea, kind of thing. [...] I'll see something and instantly recognize what problems that it solves. And so, while I didn't think of the answer in the first place, I'll think, ah, I see why they've done that and think that's an interesting solution.

The above passage suggests that trade fairs can offer important opportunities to access external forms of knowledge, and that many firms exhibiting at these events engage in searches that aim to acquire new knowledge about a product or process that is relevant to them. Yet, the findings from the qualitative analysis also suggest that the types of solutions firms generally seek are rarely far removed from their existing production contexts and pre-established knowledge foundations. The above quote speaks to this sentiment by suggesting the following: "I'll see something and instantly recognize what problems it solves". The ability to "instantly recognize" how a solution that is presented at the trade fair could be applied to a particular issue or set of problems relates to the firm's "search scope" or "how widely a firm explores new knowledge" (Katila & Ahuja, 2002, 1183). In other words, it is the firm's own experience, pre-established routines, and knowledge bases that serve as a key guide in determining what problems a promoted solution might be able to solve.

While the present study is not able to distinguish between radical and incremental forms of innovation in the search patterns of firms, it is expected that German firms will search mainly for incremental innovations, while U.S. firms will be more prone to choose radical innovations due to the particular institutional arrangements of their home context (Hall & Soskice, 2001). While this general tendency is expected, it is important to recognize that firms from different countries may engage in many different types of innovation activities; there are, for example, radical innovators in Germany as well as U.S. firms that engage in incremental innovations at times. The main point, however, is that even if a German firm, for instance, does search for a radical innovation or solution at the trade fair, that radical innovation will be adapted to the firm's distinctive production and institutional context.

Beyond the question of exploitative versus explorative search, the lengthy quotation above also speaks to the presence of "solution-marketers" at trade fairs in the sense that exhibitors come to the trade fair, in part, to present ready-made "solutions". These solutions are then discovered by other exhibiting firms at the trade fair and later linked back to specific problems in their own production or technological contexts. As suggested by the garbage-can model (Cohen et al., 1972), the traditional problem-solution relationship is reversed or turned upside-down in organized anarchies. This reversed relationship is well exemplified in the context of trade fairs.

In discussing exploitative and explorative search dimensions, it is important to note that exploitative search is not meant to be seen as synonymous with divergence pathways regarding the technological specialization of firms. Similarly, the search and incorporation of new external knowledge is not meant to be interpreted as leading towards a pathway of convergence regarding such specializations. In relation to the search literature, the basic argument of this chapter is twofold: First, exploitative search processes tend to be more widespread than explorative search processes in the context of the four trade fairs examined here. This finding supports existing empirical findings in the literature on organizational and technological search regarding the prevalence of this search type (Pavitt, 1988; Cantwell, 1989; Helfat, 1994; Stuart & Podolny, 1996; Tripsas & Gavetti, 2000; Laursen, 2012). Second, in the complex setting of the trade fair, the search for new external knowledge does not necessarily mean "a conscious effort to move away from current organizational routines and knowledge bases" (Katila & Ahuja, 2002, 1184). Instead, it tends to be guided by the pre-existing knowledge bases and corresponding institutions of the firm's production context. This point is well articulated in the following statement by the manager of an integrated lighting design firm: "We look for applications that would fit our product, for applications that are new that we haven't thought of yet, but that we could fit our product to."

These initial findings provide hints as to the potential linkages between the search behaviours of firms at trade fairs and the specific institutional conditions and dynamics of distinct capitalist varieties. These issues will be explored further in the discussion section of this chapter, and then studied in an even more systematic way in the quantitative analysis that will be presented in Chap. 6. Before moving on to these questions, however, I first turn to two additional themes that emerged from the interview results. These are: i) the experiential versus cognitive nature of search processes at trade fairs, and ii) the use of specific search heuristics at these events.

5.3.2 Experiential Versus Cognitive Search Processes

As explored in Chap. 4, the experiential-cognitive search dimension focuses not on *where* (i.e., in relation to the distance from existing knowledge terrains), but on *how* search is conducted. Recall that experiential search relies on direct feed-back from the environment to inform the future actions and interactions of firms (Cyert & March, 1963; Nelson & Winter, 1982; Hsieh et al., 2007), including the adoption of specific technological pathways or solutions. Cognitive search, on the other hand, does not involve such direct environmental feed-back; rather, it relies on representations and abstractions (Gavetti & Levinthal, 2000; Gavetti, 2012). Experiential search is, thus, characterized by processes of "learning-by-doing" (Pisano, 1994), whereas cognitive search involves processes of "learning-*before*-doing" (Pisano, 1994). While I expected that the search processes of firms at trade fairs would incorporate key elements of both search dimensions, it was anticipated that experiential processes would be dominant due to the dynamics of knowledge circulation at these events and the presence of multi-sensory informational cues, which foster a method of constant comparison and in-the-moment evaluative practices.

Although the results from the qualitative interviews provide support for both experiential and cognitive dimensions of search, as expected, there was greater evidence of experiential processes taking place within the context of the trade fairs examined. Consider, for instance, the following response from a small commercial lighting firm: "Basically [our goal is] to check out new products and to see what's new, what everyone's focus is right now. We obviously want to provide that product that people are interested in." When asked how they determine what is important, the firm suggested that "usually it depends on people's reactions." This response speaks to the nature of the direct feed-back processes at trade fairs and the subsequent incorporation of such messages into the firm's future action. In a similar vein, the following quote from a medium-sized instrumentation firm explained the search process as follows: "Yes, we walk around and take notes – whether we write them down or put them in our mind mentally. And then if one of our employees here sees something, a lot of times they'll take us to that booth and say 'hey, did you see this?'" Here, again, there is a suggestion of "learning by-doing", where potential solutions are being discovered through the course of the search itself, rather than the search process being guided by a specific search objective.

With respect to the cognitive dimension of search, there were fewer clear examples of firms involved in processes of "learning-before-doing" (Pisano, 1994) or the "off-line" evaluation of alternatives (Gavetti & Levinthal, 2000; Lopez-Vega et al., 2016). However, there was quite substantial evidence regarding the use of different search heuristics, which are closely related to the cognitive search dimension. This aspect is taken up separately in the sub-section directly below.

5.3.3 The Use of Search Heuristics at Trade Fairs

Closely related to the cognitive dimension of search are the specific heuristic shortcuts used by firms in their search processes during the trade fair. While experiential, learning-by-doing, search processes seemed more prevalent at the trade fairs examined, as discussed in Chap. 4, learning vicariously from other organizations is a specific type of heuristic search that is often used by exhibiting firms. This heuristic involves observing the actions and attendant outcomes of other organizations and then adopting or mirroring the behaviours that seem most successful (Yang et al., 2010). This search heuristic was expected to feature quite prominently in the dynamics of inter-firm search and interactions at trade fairs given the unique opportunities available to firms to observe at close range the actions and visual representations of other firms, including their strongest competitors.

Overall, the results from the qualitative analysis provide strong support for the above expectation. Indeed, the qualitative interviews provided clues that the vicarious search heuristic is fairly commonly employed by firms during the trade fair. Consider, for instance, the following statement by a medium-sized decorative lighting design firm: "Yes. Everybody is looking to make sure that they are competitive or ahead of the curve on certain products. They want to know who's got what, and they want to make sure that they are […] posturing so that they are competitive in the market. Especially now, it's very important to be competitive and have a wide range of things to offer in the market. You don't want to be a one-hit-wonder." Another firm from the lighting industry corroborated this sentiment, while suggesting in more specific terms the nature of such vicarious learning: "Predominantly, I'll look around at our various competitors to see where their products are at just to get an idea to make sure that we are still in the same playing field and that they haven't come up with a new technology or are pushing their technology in a different direction to try and move the market." This quote nicely illustrates the focus on the vicarious search heuristic – namely, on the "behaviour and associated performance outcomes" of other firms (Yang et al., 2010, 7).

Finally, as suggested by the next quotation from the representative of a digital manufacturing firm, the vicarious organizational learning heuristic was also applied in the context of technology-oriented trade fairs: "Oh yeah. I will walk around and see what people are doing, see what they have decided. Something might look different or unique. For example, over here, these signs – the signage here, we found at a show a few years ago. Somebody was doing something very similar to it and for me it was a possibility, so we put it together." What is particularly illuminating about the above statement is the modeling or imitative aspect of the search heuristic. In other words, this firm observed the behaviour of another firm and its associated successful outcome at the trade fair, and then copied that behaviour with the expectation or hope of reaping similar gains.

5.4 Linking Technological Search Patterns to Capitalist Varieties

The analysis presented in this chapter suggests that while trade fairs include important mechanisms (e.g., the temporary co-location of firms from the same industry or sector) that can potentially link firms to broader patterns of technological standardization/specialization, the search activities of firms at these events are rarely guided by a linear or rational decision-making process. Indeed, the processes of search identified in my interviews at the four North American trade fairs were much more in line with the notion of "bounded rationality" and the garbage-can model of decision-making in organized anarchies than with a rational-choice model of decision-making. Consider the notion of "problematic preferences", for instance, where "[t]he organization operates on the basis of a variety of inconsistent and ill-defined preferences" (Cohen et al., 1972, 1). The results of this study suggest that such a description can be aptly used to characterize the underlying goals of search and interaction for the majority of firms at trade fairs. The following quotation nicely summarizes this point: "Yes and no. It depends on who is the person that is out there looking. Sometimes we are looking for people to partner up with, sometimes we are looking for potential customers, and sometimes we just want to see what is out there."

As for how firms engage in search during trade fairs, both the concepts of problematic preferences and unclear technologies provide a useful contextual framing. With respect to problematic preferences, for example, Cohen et al. (1972, 1) explain that individual preferences in organized anarchies tend to be discovered through the course of action, rather than action being motivated by clearly defined goals or preferences. The results of the present empirical analysis strongly support this contention. With regards to the concept of unclear technologies, moreover, organized anarchies, such as trade fairs, are said to operate "through simple trial-and-error procedures, the residue of learning from the accidents of past experience, and pragmatic inventions of necessity" (Cohen et al., 1972, 1).

In the context of the trade fair, such "pragmatic inventions of necessity" could amount to gradual shifts in the practices of a firm, regarding, for example, the adoption of a new technology or application, or the retiring of a product that has lost its relevance in the industry or market in which the firm operates. Over time, such incremental adjustments can have a substantial impact on the development pathways of firms and even the larger technology fields or industries in which they are embedded. As such, firms participating in international trade fairs may be seen as inadvertently adopting "change strategies" (Mahoney & Thelen, 2010) that, over time, may collectively amount to a more transformative institutional shift. Further, as much as firms may be viewed as inadvertent "change agents" (Mahoney & Thelen, 2010), it is the trade fair itself that may, in fact, be the most influential "change agent". Described alternatively as "windows of opportunity" or "choice opportunities" (Cohen et al., 1972), trade fairs may serve as organized temporary catalysts for change through the assembly of the various garbage-can components. With regards to trade fairs of a truly international character, the "components" that are brought together carry with them vital pieces of experience-based knowledge regarding the institutions and practices of firms from diverse capitalist systems. As such, participation in the trade fair may play a role in broader developments regarding, for instance, processes of technological convergence or divergence.

While technological convergence and the diffusion of best practices may often be viewed as an inevitable outcome of international trade fairs, the results of the qualitative analysis suggest that the technological search processes of firms are closely related to their individual competence and knowledge bases. Indeed, while processes of non-linearity and learning-by-doing are readily apparent in the descriptions provided by firms regarding their methods of search during trade fairs, this non-linearity does not necessarily imply that the search behaviours of firms are without some kind of guiding logic. As the search literature tells us, firms generally rely quite heavily on their pre-existing technology or production contexts and associated knowledge bases when searching for solutions at the trade fair. In terms of exploitative search, these underlying knowledge bases can play a decisive role in shaping the search processes of firms at trade fairs. The qualitative analysis provides clear empirical evidence of the presence of exploitative search processes at trade fairs. Thus, despite the heterogeneity of goals and strategies that are characteristic of these events, what emerges in the complex organizational setting of the trade fair are processes that, while often unplanned and non-linear, are largely contextually based.

To the extent that the competence bases of firms vary systematically according to the national economic systems in which these firms are embedded (Hall & Soskice, 2001; Crouch, 2005), international trade fairs may support rather than undermine patterns of ongoing national specialization. While further analysis based on a longitudinal design would be needed to fully explore this latter claim, developing a better understanding of the relationship between country and search type seems a useful step towards determining whether the VoC thesis holds in the face of ongoing processes of globalization and internationalization. The quantitative analysis developed in the next chapter moves us in this direction by providing a broader empirical basis for the arguments presented in this chapter. In doing so, it sheds some light on the potential macro-level implications of inter-firm search and interactions at international trade fairs.

5.5 Conclusion

Through a qualitative approach, this chapter explored the search and information processes of firms at four leading North American trade fairs. Based on an original typology of search behaviour, it developed meaningful categories for identifying and comparing the seemingly random and often non-linear ways in which firms search for "solutions" during these events. The empirical analysis demonstrated that the choices firms make and the types of searches they engage in at trade fairs are heterogeneous and rarely follow a clear or linear trajectory. The search processes of firms at trade fairs are clearly oriented towards acquiring new knowledge – be it product-, technology-, industry-, market- or management-related knowledge. Yet, in searching for new knowledge, the interview data suggested that firms generally proceed in ways that are still more or less connected to their pre-existing production contexts and technological knowledge bases. As such, it was argued that the search processes of firms at trade fairs tend to be largely based on patterns of knowledge contextualization and path-dependence. Such patterns are expected to lead towards continued technological specialization rather than standardization. This is largely due to the complex, organized anarchy-like context of the trade fair and the fact that the mixture of participants, problems, and solutions in this environment better resembles a garbage-can process of decision-making rather than a rational-comprehensive one. As research in behavioral economics suggests, the nature of individual preference formation and change is strongly shaped by the economic context in which economic actors are embedded (Camerer et al., 2004).

This chapter focused specifically on the micro processes of technological search and the exchange of corresponding knowledge at leading business-to-business fairs in North America. Yet, the findings may have important implications regarding the role of such processes in the future development of broader national production systems. Having established a more complete picture of the nature of the search and information processes at these events, the question now becomes: how and in what ways do the search processes of firms at major international trade fairs shape processes of convergence or divergence with respect to the technological specializations of firms and the political-economic models in which they operate? To what extent do leading international trade fairs enable the transfer of specific products, technologies, and practices across diverse national political economies? And, how do trade fairs mediate the negotiation of differences between capitalist economies in terms of their technological specializations and the institutional and knowledge bases underlying them? At present, the linkages between what goes on at international trade fairs, in terms of inter-firm interactions, and the dynamics of capitalist varieties are still poorly understood. The next chapter sheds some light on these important questions.

References

Bathelt, H., & Gibson, R. (2015). Learning in "organized anarchies": The nature of technological search processes at trade fairs. *Regional Studies, 49*(6), 985–1002.

Borghini, S., Golfetto, F., & Rinallo, D. (2006). Ongoing search among industrial buyers. *Journal of Business Research, 59*, 1151–1159.

Camerer, C. F., Loewenstein, G., & Prelec, D. (2004). Neuroeconomics: Why economics needs brains. *The Scandinavian Journal of Economics, 106*(3), 555–579.

Cantwell, J. (1989). *Technological innovation and multinational corporations*. Blackwell.

Cohen, M. D., March, J. G., & Olsen, J. P. (1972). A garbage can model of organizational choice. *Administrative Science Quarterly, 17*(1), 1–25.

Crouch, C. (2005). *Capitalist diversity and change: Recombinant governance and institutional entrepreneurs*. Oxford University Press.

Cyert, R. M., & March, J. G. (1963). *A behavioral theory of the firm*. Prentice-Hall.

Gavetti, G. (2012). PERSPECTIVE – Toward a behavioral theory of strategy. *Organization Science, 23*(1), 267–285.

Gavetti, G., & Levinthal, D. A. (2000). Looking forward and looking backward: Cognitive and experimental search. *Administrative Science Quarterly, 45*(1), 113–137.

Godar, S. H., & O'Connor, P. J. (2001). Same time next year – Buyer trade show motives. *Industrial Marketing Management, 30*(1), 77–86.

Gopalakrishna, S., & Williams, J. D. (1992). Planning and performance assessment of industrial trade shows: An exploratory study. *International Journal of Research in Marketing, 9*(3), 207–224.

Hall, P. A., & Soskice, D. (2001). An introduction to varieties of capitalism. In P. A. Hall & D. Soskice (Eds.), *Varieties of capitalism: The institutional foundations of comparative advantage* (pp. 1–68). Oxford University Press.

Helfat, C. E. (1994). Evolutionary trajectories in petroleum firm R&D. *Management Science, 40*, 1720–1747.

Hsieh, C., Nickerson, J. A., & Zenger, T. R. (2007). Opportunity discovery, problem-solving and a theory of the entrepreneurial firm. *Journal of Management Studies, 44*(7), 1255–1277.

Katila, R., & Ahuja, G. (2002). Something old, something new: A longitudinal study of search behavior and new product introduction. *Academy of Management Journal, 45*(8), 1183–1194.

Laursen, K. (2012). Keep searching and you'll find: What do we know about variety creation through firms' search activities for innovation? *Industrial and Corporate Change, 21*(5), 1181–1220.

Lopez-Vega, H., Tell, F., & Vanhaverbeke, W. (2016). Where and how to search? Search paths in open innovation. *Research Policy, 45*(1), 125–136.

Mahoney, J., & Thelen, K. (2010). A theory of gradual institutional change. In J. Mahoney & K. Thelen (Eds.), *Explaining institutional change: Ambiguity, agency, and power* (pp. 1–37). Cambridge University Press.

Nelson, R. R., & Winter, S. G. (1982). *An evolutionary theory of economic change*. Harvard University Press.
Pavitt, K. (1988). International patterns of technological accumulation. In N. Hood & J.-E. Vahlne (Eds.), *Strategies in global competition* (pp. 126–157). Croom Helm.
Pisano, G. P. (1994). Knowledge, integration, and the locus of learning: An empirical analysis of process development. *Strategic Management Journal, 15*(S1), 85–100.
Sashi, C. M., & Perretty, J. (1992). Do trade shows provide value? *Industrial Marketing Management, 21*(3), 249–255.
Shoham, A. (1999). Bounded rationality, planning, standardization of international strategy, and export performance: A structural model examination. *Journal of International Marketing, 7*(2), 24–50.
Simon, H. A. (1955). A behavioral model of rational choice. *The Quarterly Journal of Economics, 69*(1), 99–118.
Simon, H. A. (1956). Rational choice and the structure of the environment. *Psychological Review, 63*(2), 129.
Simon, H. A. (1987a). Behavioral economics. In J. Eatwell, M. Milgate, & P. Newman (Eds.), *The new Palgrave: A dictionary of economics* (Vol. 1, pp. 221–225). Macmillan.
Simon, H. A. (1987b). Satisficing. In J. Eatwell, M. Milgate, & P. Newman (Eds.), *The new Palgrave: A dictionary of economics* (Vol. 1, pp. 243–245). Macmillan.
Stuart, T., & Podolny, J. (1996). Local search and the evolution of technological capabilities. *Strategic Management Journal, 17*, 21–38.
Tripsas, M., & Gavetti, G. (2000). Capabilities, cognition, and inertia: Evidence from digital imaging. *Strategic Management Journal, 21*(10/11), 1147–1161.
Yang, H., Steensma, H. K., & Phelps, C. (2010). Learning from what others have learned from you: The effects of knowledge spillovers on originating firms. *Academy of Management Journal, 53*(2), 371–389.

Chapter 6
Identifying National Patterns of Firm Behaviour at International Trade Fairs: A Quantitative Analysis

6.1 Introduction

This chapter builds on the previous chapter's findings regarding the garbage-can processes of search at trade fairs and, specifically, the argument that despite the non-linear and "boundedly rational" behaviour of firms, the searches at these events tend to be strongly guided by the pre-existing production contexts and knowledge bases of the countries in which firms are embedded. Moving from the micro- to the macro-level, this chapter broadens the empirical foundations of these findings through a quantitative study of cross-national differences in firm behaviour at trade fairs. Taken together, the findings from both chapters provide some empirical support for the claim that international trade fairs have become key drivers of global economic dynamics, mediating the negotiation of differences between capitalist economies regarding their technological specializations and the institutional practices and knowledge bases that underpin them. Specifically, this chapter demonstrates that not only are the search processes of firms largely path-dependent, but also that firms from different countries are embedded in distinct paths and tend to follow their specific pathway.

The overall expectation in this chapter is, thus, that firms from distinct capitalist varieties will exhibit observable differences in behaviour, as their search processes and characteristics are fundamentally shaped by, and embedded within, the institutional conditions and corresponding system logics that define these varieties. While this is the main hypothesis, it is important to note that the results may not be entirely conclusive on this front. This is due to the heterogeneity of the actors involved as well as the complexity of the processes that occur during trade fairs. Multi-national firms, for example, often play a significant role at international trade fairs. Interviewing these types of firms may lead to more muddled results, as one might argue that multi-national firms will adjust their structure and characteristics to the

R. Gibson, *International Trade Fairs and Inter-Firm Knowledge Flows*,
https://doi.org/10.1007/978-3-031-20557-6_6

different country contexts in which they work. To the extent that this is true, considerable heterogeneity might be expected. Moreover, the questions asked of firms in this study were necessarily quite concise and broadly defined. As such, they cannot fully capture the context-specific insights from the firms that are needed to make sense of such heterogeneities. Nevertheless, it is hoped that the tendencies revealed will lend support to the main argument developed in this research.

In order to test whether the actions and interactions of firms at trade fairs differ in ways that correspond to the distinct capitalist models in which they operate, this chapter investigates general similarities and differences between German and U.S. firms along a variety of dimensions related to the exchange of knowledge at trade fairs. Drawing on the VoC literature, the analysis below is driven by such questions as the following: (i) Do U.S. firms rely more heavily than German firms on online sources of information to develop technological and industry know-how? (ii) Do firms from Germany and the U.S. acquire knowledge from different sources during the trade fair (e.g., suppliers, competitors)? (iii) Are German firms more inclined towards customized and high-quality production in their search patterns than U.S. firms? (iv) Do U.S. firms place greater emphasis than German firms on low-cost production and constantly changing product lines in their search patterns? (v) When searching for new business partners, do both U.S. and German firms look for partners that possess core competencies that are similar or different from their own?

The quantitative analysis focuses specifically on German and U.S. firms because these countries are widely considered ideal-typical cases of coordinated and liberal market economies, respectively, in the VoC literature. The analysis is based on a total of 165 surveys from firms that previously exhibited at major international trade fairs, 60 of which are German and 105 of which are American. The analysis includes a comparison of the average evaluations of firms on questions regarding the use of specific technological information sources, the core capabilities that are central to the operations of the firm, as well as the core capabilities the firm looks for in potential business partners. Because I found that U.S. firms generally placed more value than German firms on information obtained through trade fairs,[1] it was useful to also include a measure of the relative evaluations (RE) of firms. As will be discussed in more detail later on, the RE is essentially the difference between a country's average evaluation on a specific item (e.g., strong R&D capabilities) and the average evaluation of that country across all items. Both the average and relative evaluations are based on a 7-point Likert scale. Statistical tests, including a series of t-tests and multivariate regression, were used to identify significant differences between German and U.S. firms with respect to how they act and interact during the trade fair. While the data were generated at the micro-level and do not provide a sample large enough or representative enough to speak conclusively about macro-level outcomes, they provide sufficient grounds upon which to draw careful conclusions regarding the broader country-level implications of inter-firm interaction during trade fairs.

[1]This trend is consistent with the VoC thesis and relates to the greater emphasis placed by CMEs than LMEs on ongoing relationships as a key source of information.

The next section begins by exploring the following question: Do U.S. and German firms differ with respect to the types of activities they engage in, or the information sources they use, to learn about new technological and market trends in their industry? By focusing on the micro processes of firm behaviour during trade fairs, this section identifies patterns of difference that may be linked to the broader institutional configurations that define different capitalist models. This helps connect the findings from the previous chapter regarding the nature and dynamics of firms' searches at trade fairs to the specific institutional arrangements associated with LMEs and CMEs.

The second section of the chapter moves beyond basic information sources to look more closely at the character and pathways of knowledge flows at trade fairs. My primary interest, here, is to determine whether there are significant differences between German and U.S. firms with regards to the following: (i) where they get knowledge at the trade fair, (ii) to whom they provide knowledge, (iii) what types of knowledge they acquire at the trade fair, and (iv) what types of knowledge they provide to others at these events.

The third section focuses on the core competencies of German and U.S. firms, as well as the core competencies of potential business partners. This is important not only to determine whether there are broad differences between the two capitalist varieties, but also to identify whether or not firms search for competencies that match their existing strengths. The larger objective here is to situate such observations within the context of discussions regarding the persistence or decline of distinct national specializations – that is, the convergence-divergence debate (Crouch & Streeck, 1997; Kitschelt et al., 1999; Yamamura & Streeck, 2003), as well as recent discussions regarding the dynamic elements of capitalist diversity (Deeg & Jackson, 2007; Jackson & Deeg, 2008), including incremental forms of institutional change (Thelen, 2004; Streeck & Thelen, 2005; Hall & Thelen, 2009; Mahoney & Thelen, 2010).

The fourth section builds on these findings by exploring the actual degree to which the core competencies of firms match those of their potential partners. This section relies on an original homogeneity/complementary scale (see Chap. 1), which measures the match between the competencies of firms and those of the partners they look for at trade fairs. In this scale, '0' represents complete homogeneity and '6' represents the maximum amount of distance between the firm's own central capabilities and those it looks for in its partners. The last section concludes with a summary of the chapter's main findings.

6.2 Cross-National Differences in Industry and Technological Information Systems

As discussed in Chap. 3, the VoC literature suggests that there are fundamental differences in the nature of firm behaviour and interactions across countries, and that these are rooted in the unique institutional configurations that characterize different models of capitalism. In CMEs like Germany, for example, non-market relations prevail and there are incentives to foster more collaborative and enduring relations between firms (Hall & Soskice, 2001; Hancké, 2009). In contrast, LMEs, such as the U.S., are characterized by more competitive, arms-length relations between firms, formal contracting, and weakly regulated product markets (Hall & Soskice, 2001; Hall & Gingerich, 2004; Hancké, 2009). The existence of these distinct institutional arrangements has important implications for the organization of production (Kitschelt, 1991; Streeck, 1996; Deeg, 1999) and the nature of economic and technological specializations (Hall & Soskice, 2001; Boschma & Capone, 2015). Of course, in theory, firms have the freedom to choose their own product market strategies; yet, because such strategies are closely tied to different requirements – specific forms of relations to owners, for example, or relations with other firms – some strategies are better supported than others by the institutional arrangements of a given country (Deeg, 1999).

In CMEs, for instance, where firms depend heavily on long-term, collaborative, non-market-based relationships, economic specializations tend to occur largely in sectors and product areas where incremental forms of innovation are dominant. These products or sectors typically include: machinery, electrical equipment, chemistry, drugs, transportation, and weapons (Akkermans et al., 2009; Boschma & Capone, 2015). This is because CMEs are defined by specific or co-specific assets – that is, "assets that cannot readily be turned to another purpose and assets whose returns depend heavily on the active cooperation of others" (Hall & Soskice, 2001, 17; see also, Hall & Gingerich, 2004).

In contrast, the economic activities of firms in LMEs are organized through hierarchies and market relations, which tend to encourage specializations in sectors and products where radical innovations prevail. Included in these sectors or product areas are: electronics, aerospace, telecommunications, biotechnology, science instruments, agriculture, and food (Akkermans et al., 2009; Boschma & Capone, 2015). This is due to a larger investment in generic assets or switchable assets – that is, "assets whose value can be realized if diverted to other purposes" (Hall & Soskice, 2001, 17; Hall & Gingerich, 2004).

Given such differences, it is expected that German and U.S. firms will differ in significant ways with respect to the types of tools or information sources they use to develop technological and industry knowledge or "know-how". Firms from the U.S., for example, might be expected to rely more heavily on online communication tools and information sources, such as general social media (Twitter, Instagram, blogs), given that the development of rapid and abrupt innovations may need to be fed by fast and continuous flows of the latest cutting-edge information. Such high-speed flows of online information may not be as pertinent, however, in supporting the more incremental forms of innovation that are characteristic of CMEs like Germany. Instead, German firms might depend more on tools that facilitate and strengthen close partnerships in, for example, research and development. For German firms, participating in professional organizations or industry associations may, thus, be preferred ways of acquiring relevant industry or technological information.

While information sources are not a key focus of this study, they are a useful starting point in exploring differences between countries in patterns of firm behaviour and interactions that are connected to broader processes of technological innovation and knowledge exchange. In some instances, such information sources may not be that far removed from the search processes of firms discussed in the previous chapter. The centrality of online information sources to a firm's business strategy may, for instance, suggest a greater tendency towards cognitive search processes and learning-*before*-doing (Pisano, 1994; Gavetti & Levinthal, 2000) as opposed to experiential search and learning-by-doing (Pisano, 1994; Gavetti & Levinthal, 2000). For these reasons, the quantitative portion of the analysis begins with an exploration of the differences in information sources prioritized by firms from the U.S. and Germany.

6.2.1 General Results

To test the hypothesis that German and U.S. firms will differ in terms of the tools and information sources they consider most valuable in the development of industry and technological know-how, I used t-tests to compare differences in their mean evaluations of ten specific information sources. For this question, firms were asked to rate the importance of the individual information sources based on a 7-point Likert scale, where 1 = "least important" and 7 = "most important".[2] Table 6.1 below presents the results, with the third column highlighting the degree of difference between the average evaluation scores of the U.S. and Germany and how significant these differences are. Of the ten different information sources included in the analysis, there were significant differences between the two countries with respect to four of these sources: (i) use of specific industry websites, (ii) use of general social media, (iii) sharing professional information or advice over the telephone, and (iv) having lunch meetings with other firms or organizations.

Table 6.1 reveals some interesting and significant differences between Germany and the U.S., particularly with respect to the use of online platforms for the development of industry and technological knowledge or know-how. The difference between the two countries is most striking with respect to the use of general social media (e.g., Twitter, Facebook, blogs), but it is also seen in firms' reliance on specific industry websites as information sources. In the case of general social media, for instance, there are nearly 2 full points of difference between the two countries,

[2] There is considerable and ongoing debate in the literature regarding what constitutes the optimal number of response options in Likert scales (see, for example, Lee & Soutar, 2010; Dillman et al., 2009; Fink, 1995; Foddy, 1994). That being said, the 5-point and 7-point scales tend to be the most common (Pearse, 2011; Dawes, 2008). Whereas the 7-point scale usually allows options at the extreme ends (e.g., 1 = very strongly agree; 7 = very strongly disagree), 5-point scales typically omit these extremes (Pearse, 2011). In this study, I tried to capture a finer level of difference in preferences among firms and, thus, used the 7-point scale.

Table 6.1 Average evaluations of technological information sources used by firms from the U.S. and Germany

	U.S.	Germany	Δ
Information sources			
Industry/association events	4.99	4.55	0.44
Specific industry websites	5.23	4.71	0.52**
General social media (e.g., blogs, Twitter)	4.22	2.32	1.90***
International trade fairs	5.47	5.61	−0.14
Professional information/advice over phone	4.63	3.98	0.65**
Lunch meetings w/other firms/organizations	3.71	3.23	0.48*
Professional organizations/associations	4.12	3.79	0.33
N	99	58	

Notes: T-tests were used to identify and compare differences in the average evaluations of firms. These evaluations are based on a 7-point Likert scale, where 1 = "least important" and 7 = "most important

*$p < 0.10$ **$p < 0.05$ ***$p < 0.01$

based on a 7-point Likert scale. This finding is significant at the 1% level. In a similar vein, the perceived utility of obtaining information from specific industry websites is also considered more important by U.S. than German firms. While this finding is significant at the 5% level, the degree of difference between the two countries on this information source is only half a point on a 7-point Likert scale. Beyond Internet-based information sources, the only significant differences between the two countries involve sharing professional advice or information over the telephone and participating in lunch meetings with other firms or organizations. The former is significant at the 5% level, while participating in lunch meetings is significant at a more modest 10% level, revealing less than half a point difference in the ratings of U.S. and German firms based on the 7-point Likert scale.

Thus, the most striking finding from Table 6.1 is the difference between countries regarding their use of Internet-based information sources, with online platforms being viewed as significantly less important by German firms compared to their U.S. counterparts. Although not directly related to the study's core research questions, this finding seems consistent with the VoC approach. The institutional conditions of LMEs like the U.S., for example, favour switchable assets and promote radical forms of innovation (Hall & Soskice, 2001). In such a context, firms might well be expected to make use of a wider array of information inputs, including the high-speed flow of Internet-based information on industry trends and technological developments occurring globally. The immediacy of informational messages circulated through general social media, for example, seems to correspond to the rapid innovation style that characterizes LMEs. In contrast, it is not surprising that high-speed online information sources are not a strategic focus for firms from Germany. Due to the particular institutional setting of CMEs, the emphasis for German firms tends to be more on high quality and customized production achieved through incremental innovations and adjustments (Hall & Soskice, 2001). As such, these firms might be expected to focus on just a few trusted sources of information based on long-term collaborative relationships.

6.2.2 *Findings Related to Firm Age, Size, and Sector*

Table 6.1 provides a preliminary picture of differences in the information sources used or preferred by German and U.S. firms; however, the data presented are fairly rough in the sense that they do not take into account other potentially influential characteristics of the firm, such as its size, age, and the sector in which it operates. As such, Table 6.2 further distinguishes the significant sources of information identified above based on three additional criteria that operate as control variables: age, size, and sector. With respect to the age of the firm, I defined two categories for the analysis: (i) *post-globalization* firms, which are firms that were established in 1990 or later, corresponding to the start of the globalization era; and (ii) *pre-globalization* firms, which are firms that were established prior to 1990, or before the onset of the globalization period. In terms of firm size, I again defined two categories: (i) *small* and (ii) *medium-sized or large* (ML) firms. Small firms are those that have 100 or fewer employees, whereas ML firms are those with more than 100 employees. Finally, the two sectors compared in the analysis were the *electronics sector* and the *machinery sector*.

As Table 6.2 illustrates, firms operating in the electronics sector tended to place much more importance on online sources of information than did firms from the machinery sector. While this was not apparent in the case of specific industry websites, it was highly significant in the case of general social media (e.g., blogs, Twitter). There is, for example, nearly a full two-point difference between electronics firms and machinery firms with respect to their use of general social media as an information tool to develop industry and technological know-how. This finding is significant at the 1% level. Although firm size does not seem to matter with respect to the information sources used to develop industry and technological learnings, age does make a difference. As Table 6.2 suggests, firms established after the onset of the globalization period – referred to here as "post-globalization" firms – were more likely than their "pre-globalization" counterparts to rely on general social media. These findings are also significant at the 1% level.

Thus, generally speaking, firms from the electronics sector and post-globalization firms revealed a clear tendency towards the use of online information platforms to develop industry and technological know-how. At an intuitive level, these findings are perhaps not surprising. The electronics sector is typically characterized by radical forms of innovation (Akkermans et al., 2009; Boschma & Capone, 2015). As discussed above in relation to the importance of online information for U.S. firms, the development of radical forms of innovation may depend to a greater extent on high-speed online information flows and near instantaneous communication than do incremental forms of innovation. As for age, post-globalization firms in the early stages of business development often have fewer financial and human resources than firms established before the onset of the globalization period (Aldrich & Auster, 1986; Carayannopoulos, 2009; Aarstad et al., 2015). Post-globalization firms also generally have smaller, less diverse networks compared to their more experienced, pre-globalization counterparts (Hite & Hesterly, 2001; Aarstad et al.,

Table 6.2 Average evaluations of the importance of information sources used by firms, by age, size, and sector

	Age			Size			Sector		
	[a] Post-globalization	[b] Pre-globalization	Δ	[c] Small	[d] ML	Δ	Electronics	Machinery	Δ
Activities									
Specific industry websites	5.09	4.91	0.18	5.08	4.96	0.12	5.13	4.88	0.25
General social media (e.g., blogs, Twitter)	3.98	3.11	0.87***	3.54	3.31	0.23	4.20	2.38	1.82***
Professional info/ advice over phone	4.60	4.29	0.31	4.42	4.31	0.11	4.52	4.18	0.34
N	57	92		92	50		98	59	

Notes: T-tests were used to identify and compare differences in the average evaluations of firms. These evaluations are based on a 7-point Likert scale, where 1 = "least important" and 7 = "most important"

$^{*}p < 0.10$ $^{**}p < 0.05$ $^{***}p < 0.01$

[a]Post-Globalization refers to firms that were established in 1990 or later, corresponding with the start of the globalization period

[b]Pre-Globalization refers to firms that were established after 1990

[c]Small firms are those that have 100 or fewer employees

[d]ML denotes Medium-sized and Large firms. In this study, ML firms are those that have more than 100 employees

2015). For these firms, online information sources may be especially helpful, as they are virtually free and most can be accessed at almost any time of day and in any location (e.g., home-based office). Though partial in nature, online information sources also provide quick and easy access to global information related to the firm's specific technological or product focus.

For the reasons outlined above, the greater importance placed on online information and communication platforms by post-globalization firms and electronics firms seems to make sense. The critical question that arises is whether the independent effects of firm age and sector will undercut the previously observed significance of country in explaining differences in firms' choices of information sources. This question will be investigated more thoroughly later on through multivariate linear regression analysis.

6.2.3 Results of Additional Control Variables

In addition to firm age, size, and sector, it is useful to further distinguish among other firm characteristics and behaviours that may play a role in the information sources firms use to develop industry or technological know-how. To this end, the following five firm characteristics and behaviours were included as control variables: (i) the international versus domestic focus of the firm's operations; (ii) whether or not the firm engages in R&D collaborations with other firms; (iii) whether or not the firm has merged with or acquired another firm in the past 3 years; (iv) whether or not the firm co-operates with external organizations in industry standard-setting; and (v) whether or not the firm has received government support to participate in the trade fair.

While Table 6.3 below reveals a number of significant relationships between the selected information sources and the specific characteristics of firms, the most striking area of difference is between the international or domestic focus of a firm's operations and its use of online information platforms. More specifically, firms with a primarily domestic focus of operations had a tendency to view the use of general social media as more important than firms with a mainly international focus of operations. This finding is significant at the 1% level. One explanation for this difference might be that firms with mainly domestic operations have fewer opportunities to access global knowledge affecting their industry through, for example, face-to-face meetings and knowledge exchanges with international partners. They may, therefore, turn to online information flows and communication platforms as a way of overcoming this shortcoming.

A second general trend is that firms that had not received government support to participate in the trade fair tended to place higher importance on the use of online information sources than those who had received government support. In the case of general social media, this finding is significant at the 5% level. The reasons for this are not clear. One possibility is that there is a specific type of firm that receives government support for trade fair participation – perhaps firms that are newer and

Table 6.3 Average evaluations of information sources used by firms, by specific firm functions and behaviours

	[a]Internationalization			[b]R&D Collaborations			[c]Merged or acquired			[d]Industry standard-setting			[e]Received government support		
	International	Domestic	Δ	No	Yes	Δ	No	Yes	Δ	No Participation	Participate	Δ	No	Yes	Δ
Activities															
Specific industry websites	5.09	5.00	0.09	4.93	5.10	−0.17	4.98	5.17	−0.19	4.78	5.25	−0.47*	5.04	5.12	−0.08
General social media	3.08	4.00	−0.92***	3.32	3.75	−0.43	3.72	3.09	0.63*	3.39	3.59	−0.20	3.67	2.84	0.83**
Professional information/advice over the phone	4.54	4.29	0.25	4.20	4.60	−0.40	4.30	4.61	−0.31	4.19	4.62	−0.43	4.47	4.04	0.43
N	79	78		73	81		110	47		74	79		122	25	

Notes: T-tests were used to identify and compare differences in the average evaluations of firms. These evaluations are based on a 7-point Likert scale, where 1 = "least important" and 7 = "most important"

$^{*}p < 0.10$ $^{**}p < 0.05$ $^{***}p < 0.01$

[a]Internationalization refers to whether the operations of the firm are mainly international or domestic

[b]This refers to whether the firm engages in R&D collaborations with other firms

[c]Has the firm merged with or acquired another firm in the past 3 years?

[d]Refers to whether the firm co-operates with external organizations in industry standard setting

[e]Has the firm ever used government support to participate in trade fairs?

have fairly modest revenues. For reasons aside from the variable of government support itself, this general "type" of firm may be drawn to online information sources because of their relative accessibility, low cost, and so on. Finally, while the results were not significant, the table revealed a general tendency that the additional firm characteristics described above had no real relationship with the use of specific industry websites or obtaining professional information/advice over the phone.

As with Table 6.2, the inclusion of additional control variables seems to generally support, or at least not weaken, the chapter's original findings regarding the distinction between German and U.S. firms with respect to their behaviour at international trade fairs. The extent to which the country difference holds in the face of a variety of control variables will be further investigated and tested through multivariate linear regression analysis later on in this chapter.

6.2.4 *Regression Results*

The findings presented thus far provide some initial evidence that the country in which a firm operates may play an important role in shaping the behaviour of firms at trade fairs, specifically with respect to the types of information sources firms' use or the types of activities they engage in to develop industry and technological know-how. The question then becomes: Do the country differences noted above persist when one considers the independent effects of the other relevant control variables (e.g., size, age, sectoral focus, international/domestic orientation) on the information sources of firms surveyed at trade fairs? In order to refine our understanding of the relationships between the firm's nationality and the information sources it uses to acquire industry and technological knowledge, I conducted multivariate linear regression analyses (SPSS Statistics 24).

Table 6.4 below includes as dependent variables the three information sources that showed significant differences between the U.S. and Germany in the initial comparison of means test (see Table 6.1). These are: *(i) General Social Media; (ii) Specific Industry Websites;* and *(iii) Professional Advice/Information over the Phone.* The independent variables are the specific characteristics or behaviours of firms explored above in Tables 6.2 and 6.3. In addition to these factors, I also included a variable on the survey distribution mode in order to take into account any possible biases that might arise from completing the survey through different formats – namely, Internet vs. Paper. My expectation is that country will remain a significant factor influencing the choice of information sources even after taking into account the independent effects of these other variables.

This type of analysis is useful in illuminating parts of the larger research puzzle regarding the nature of inter-firm interaction processes at trade fairs and the role of such interactions in shaping patterns of technological specialization. One could argue, for example, that when firms internationalize they are likely to adapt their respective models and let go of their original identity, which was based on the institutional conditions and arrangements of the firm's home country. Such a finding

Table 6.4 Regression analysis: factors influencing the firms' use of technological information sources

	General social media	Specific professional websites	Professional information via telephone
Country			
Germany vs. U.S.	1.05 (0.42)**	0.44 (0.38)	0.73 (0.40)*
Firm characteristics			
Trade fair type			
Machinery vs. electronics	0.87 (0.44)**	−0.22 (0.40)	0.23 (0.42)
Size of firm			
Small vs. ML	0.50 (0.34)	−0.08 (0.31)	−0.13 (0.32)
Age of firm			
Pre-globalization vs. post-globalization	0.29 (0.34)	0.22 (0.31)	0.05 (0.33)
Firm behaviour			
International vs. domestic	0.21 (0.34)	−0.16 (0.31)	−0.51 (0.33)
Engaged in R&D	−0.14 (0.33)	0.03 (0.30)	0.21 (0.32)
Merged in past 3 years	−0.55 (0.33)*	0.20 (0.30)	0.23 (0.32)
Participates in industry Standard-setting	0.25 (0.33)	0.49 (0.30)	0.45 (0.32)
Received government support	−0.55 (0.41)	0.15 (0.37)	−0.21 (0.39)
Survey distribution mode			
Internet vs. paper	0.42 (0.33)	−0.22 (0.30)	0.45 (0.32)
R-squared	0.29	0.06	0.12
N	125	126	125

Dependent Variables coded on a 7-point scale, with 1 = "least important" and 7 = "most important"
ML denotes Medium-sized and Large firms
The independent and control variables are coded as follows: U.S. = '1', Germany = '0'; Machinery = '0', Electronics = '1'; Small firm = '0', ML firm = '1'; Pre-globalization firm = '0', post-globalization firm = '1'; International focus of operations = '0', Domestic focus = '1'; "yes" engaged in R&D = '0', "no" did not engage in R&D = '1'; "yes" merged with/acquired another firm in the last 3 years = '0', "no" did not merge with another firm in the last 3 years = '1'; "yes" participates in industry standard-setting '0', 'no" does not participate in industry standard-setting '1'; "yes" received government support to participate in the trade fair = '0', "no" did not receive government support = '1'; Internet survey distribution = '0', Paper, F2F distribution = '1'
Notes: Unstandardized coefficients presented in cells with standard errors in parentheses
$^{*}p < 0.10$ $^{**}p < 0.05$ $^{***}p < 0.01$

would constitute a significant challenge to the argument presented in this study. Yet, as the regression analysis in Table 6.4 demonstrates, the internationalization variable had no significant impact on the information sources firms used to develop industry and technological know-how. Still, it is necessary to be cautious when making overarching statements regarding any observed relationships, given that the sample is both relatively small and may not be fully representative of the larger population of firms that exhibit at trade fairs due to the nature of the data collection.

It is also important to note that the fairly concise questionnaire used in this study (see Appendix A) cannot capture all of the variables that may have an impact on the dependent variables explored below. This is partially demonstrated by the low R-squared score.

Table 6.4 draws attention to at least two important findings regarding the information sources used by firms to develop industry and technological know-how. First, the analysis shows that country differences persist in the face of other control variables. U.S. firms were significantly more likely than their German counterparts to utilize general social media (e.g., blogs, Twitter, Facebook) as a means to acquiring knowledge about new technologies and developments in their industry. This difference is intriguing and may stem from the specific institutional arrangements that support the coordination of economic action and interaction in liberal versus coordinated market economies. It has been noted, for example, that the German way of doing business is "less instantaneous" than in some other countries, such as the U.K. or the U.S., both of which are classified as LMEs. In the German context, business relationships that are "developed slowly over time are seen as a more desirable way to do business" (Coleman, 2014). This observation fits well with key expectations derived from the VoC approach. The institutional structures of CMEs, such as Germany, promote direct and repeated interactions among firms and the development of long-term business relationships (Hall & Soskice, 2001; Hancké, 2009).

Aside from the nationality of the firm, a second important result from Table 6.4 has to do with the sectoral focus of the firm. Recall that sector was a significant factor in the bivariate analysis presented in Table 6.2. It remained significant for general social media, but was not significantly related to the other two technological information sources – namely, the use of specific professional websites and professional information or advice over the telephone. A further finding from the multivariate analysis was that survey distribution mode had no significant impact on the information sources used by firms to develop industry and technological know-how.

6.3 Cross-National Patterns of Knowledge Flows

Having developed a better understanding of the differences between German and U.S. firms with respect to the information sources they use to learn about new technological and industry trends, this section turns to look more closely at differences between the two countries regarding the nature of knowledge flows at trade fairs. More specifically, it first looks at where firms get knowledge during the trade fair, as well as to whom they provide knowledge. Second, it looks at the categories or types of knowledge firms acquire from these sources, as well as the types of knowledge they tend to provide to others. These patterns of knowledge flows at trade fairs are explored through a series of t-tests comparing firms' average evaluations of the importance of different knowledge dimensions. While the discussion in Chap. 5 revealed important insights into the micro processes of search and knowledge flows at trade fairs, this chapter aims to provide a broader basis for such findings by providing empirical evidence on cross-national patterns of knowledge flows at these events.

Table 6.5 below explores the relationships between U.S. and German firms with respect to the following questions: (i) Where does your firm get knowledge during the trade fair? (ii) To whom does your firm provide knowledge? In responding to these questions, firms were asked to rate six different knowledge sources/recipients based on the same 7-point Likert scale used in the previous analyses. These knowledge sources are: (i) existing customers; (ii) potential customers; (iii) competitors; (iv) existing suppliers; (v) potential suppliers; and (vi) other partners in research or distribution. T-tests were used to identify and compare differences in the average evaluations of firms from the U.S. and Germany across these variables. It is expected that customers will be an important source of knowledge for both U.S. and German firms; however, there may be differences between the two countries with respect to the value placed on competitors and suppliers as key sources of information at the trade fair. Based on the VoC literature, one might anticipate firms from CMEs to have greater incentives to develop enduring relationships with suppliers, for example, whereas relationships with suppliers may be more short-lived for firms operating in LMEs due to their more rapidly shifting needs.

Table 6.5 demonstrates a significant difference between firms from the two countries with respect to where firms acquire knowledge during the trade fair. U.S. firms tended to view both existing and potential suppliers as more important sources of knowledge than did German firms. There were no significant differences between the two countries with respect to the role of customers or competitors as valuable sources of knowledge at the trade fair. While customers were unsurprisingly rated as the most important source of information at the trade fair, competitors were also

Table 6.5 Sources of firms' knowledge acquisition and recipients of firms' knowledge, by country

	Where the firm gets knowledge			To whom the firm provides knowledge		
	United States	Germany	Δ	United States	Germany	Δ
Existing customers	5.76	5.71	0.05	6.17	5.98	0.19
Potential customers	5.71	5.43	0.28	6.31	5.66	0.65***
Competitors	4.80	4.50	0.30	2.70	2.62	0.08
Existing suppliers	4.95	4.47	0.48*	4.77	4.21	0.56**
Potential suppliers	4.82	4.25	0.57**	4.71	3.77	0.94***
Other Partners in Research/ distribution	4.65	4.44	0.21	4.50	4.06	0.44
Total	104	59		104	57	

Cell entries are mean scores

Dependent variables are 1–7 Likert scales, ranging from 1 = "Least Important" to 7 = "Most Important"

*p < 0.10 **p < 0.05 ***p < 0.01

rated reasonably highly by both U.S. and German firms (M = 4.80 and M = 4.50, respectively, on a 7-point scale). This may be due to the unique opportunities provided by trade fairs to observe the latest innovations of competitors, their visual representations, their actions and interactions with customers, suppliers, and so on. The emphasis on competitors as a key source of knowledge may also speak to a type of heuristic search discussed in Chap. 5, which involves learning vicariously from other firms, particularly competitors (Yang et al., 2010).

With respect to the most important recipients of firms' knowledge at trade fairs, there were significant differences between the two countries in two areas – customers and suppliers. Compared to their German counterparts, U.S. firms placed more importance on potential customers as beneficiaries of knowledge during the trade fair. This finding is significant at the 1% level. Whereas both U.S. and German firms evaluated the importance of customers quite highly on this question, it is with respect to potential as opposed to existing customers that the major difference was found. The greater importance placed by U.S. firms on potential customers may be interpreted in light of the more competitive nature of inter-firm relations in LMEs compared to CMEs, and possibly relates to the greater emphasis U.S. firms may place on looking for partners in sales within the trade fair context.

In addition to customers, U.S. firms placed greater importance than their German counterparts on both existing and potential suppliers as recipients of their knowledge. The former is significant at the 5% level, whereas the latter is significant at the 1% level. Comparing both the sources of knowledge for firms and the recipients of firms' knowledge during trade fairs, there was, thus, a clear and significant difference between the two countries regarding the emphasis placed on suppliers. Knowledge flows to and from suppliers at trade fairs were deemed significantly more important for U.S. firms than they were for German firms. This difference might be, at least, partially due to the specific institutional arrangements that shape economic activity in the different capitalist models. As the VoC literature suggests, CMEs like Germany tend to encourage firms to maintain long-term relationships with their suppliers, whereas firms in LMEs like the U.S. are more inclined to choose their suppliers in response to their immediate and continuously shifting needs (Hall & Soskice, 2001). Focusing on *potential* suppliers during the trade fair may, therefore, be less relevant for German firms, as it is actually the existing, long-term supplier relationships that are most pertinent to their operations.

In addition to exploring the knowledge flows to and from exhibiting firms during trade fairs, I am also interested in developing a better understanding of the specific types of knowledge firms consider important and whether there are significant differences in these knowledge types between countries. This is important because it relates to the multiplicity of knowledge flows running through the trade fair and the garbage-can-like processes of organizational choice that feature so prominently in such settings. While I expected that the majority of firms, regardless of their nationality, would be highly interested in the two-way exchange of product-related and technology-related knowledge at the trade fair, I anticipated that firms from the U.S. would place greater importance on market-related knowledge at these events. This is because in LMEs like the U.S. the market plays a leading role in coordinating economic action, whereas, in CMEs, there is much more state involvement in coordinating such activity (Hall & Soskice, 2001; Hall & Gingerich, 2004; Hancké, 2009).

As Table 6.6 below demonstrates, there were clear and significant differences between U.S. and German firms with respect to market-related knowledge. More specifically, U.S. firms placed more importance on acquiring market-related knowledge at the trade fair than their German counterparts. This was true both in terms of the knowledge the firm acquired during the trade fair as well as the knowledge it provided to others at these events. This finding makes sense in relation to the observed differences in capitalist varieties, specifically regarding the larger role of market forces in the coordination of economic activity in LMEs (Hall & Soskice, 2001; Hancké, 2009). This is not to suggest that market-related knowledge is unimportant from the perspective of German firms – indeed, the mean evaluation score of 5.09 on the 1–7 scale suggests that such knowledge is important to German firms (See Table 6.6). However, differences in the degree of state involvement in the economies of the two countries may provide at least a partial explanation of why U.S. firms were generally more preoccupied with obtaining market-related knowledge at trade fairs than were German firms.

In addition to market-related knowledge, Table 6.6 points to a difference between the two countries with regards to industry-related knowledge. More specifically, U.S. firms placed greater weight on acquiring this type of knowledge at trade fairs than did German firms. This finding is significant at the 1% level. One potential explanation for this difference might involve differences in the nature of industry specializations and the organization of decision-making in the two distinct political economies. As previously noted, industry specializations in the two countries tend to correspond to cross-national differences related to radical versus incremental forms of innovation, with the U.S. enjoying particular strengths in the former and Germany having comparative advantages in the latter (Akkermans et al., 2009; Boschma & Capone, 2015). While the knowledge acquired at trade fairs may be relevant to processes of incremental innovation, firms in CMEs also tend to rely heavily on dense networks of public and semi-public institutions and research institutes. Defined by consensus-building and more collaborative forms

Table 6.6 Types of knowledge acquired and provided by firms, by country

	Knowledge firm gets			Knowledge firm provides		
	United States	Germany	Δ	United States	Germany	Δ
Product-related	5.80	5.59	0.21	6.20	6.05	0.15
Technology-related	5.64	5.31	0.33	5.57	5.48	0.09
Market-related	5.83	5.09	0.74***	5.00	4.33	0.67***
Industry-related	5.71	5.02	0.69***	4.95	4.61	0.34
Management-related	3.98	3.93	0.05	3.41	3.46	−0.05
Total	104	59		104	58	

Cell entries are mean scores
Dependent variables are 1–7 scales, ranging from 1 = "Least Important" to 7 = "Most Important"
*$p < 0.10$ **$p < 0.05$ ***$p < 0.01$

of decision-making, these networks are instrumental in providing insight into industry-related problems as well as arriving at industry-wide solutions. As such, trade fairs may be viewed as a supplementary avenue through which CMEs develop industry-related knowledge, whereas the lack of similar networks in LMEs might help explain why U.S. firms place relatively greater weight on this knowledge resource.[3]

6.4 The Core Capabilities of Firms and Potential Partners

The discussion thus far points to some clear differences regarding the importance placed on specific information sources and knowledge processes by German and U.S. firms at trade fairs. The question now becomes, can these cross-national differences be further extended to the character and search behaviours of firms? For example, do firms from the U.S. and Germany reveal patterned differences with regards to their core competencies? Moreover, when it comes to seeking out new business alliances, do German and U.S. firms look for partners that possess core capabilities that are largely similar or different from their own? These questions are important not only in establishing whether there are, in fact, cross-national differences in the specializations of firms exhibiting at trade fairs, but also in determining whether the partners being sought through these events are related or unrelated to their existing specializations or competencies. To the extent that potential partners match the core competencies of firms, there may be some evidence to support the hypothesis that the future development pathways of such firms will reinforce rather than dilute existing national specializations, thereby leading toward processes of continued divergence rather than convergence.

Before addressing these questions, however, it is first necessary to determine whether firms do, in fact, look for partners during trade fairs, rather than simply using these events to promote their own products and generate new sales or leads. The findings presented in the table below support recent studies that emphasize the learning and networking functions of international trade fairs over the purely sales and promotional functions of these events (see, for example, Rinallo & Golfetto, 2011; Bathelt et al., 2014). As Table 6.7 shows, the vast majority of both U.S. and German firms indicated that they look for potential partners during the trade fair, at 90% and 85%, respectively. There were no significant differences between the two

[3] It is important to stress that the purpose of this study is not to demonstrate the importance of trade fairs vis-à-vis other potential information or knowledge sources. The objective of this chapter is, instead, to analyze and compare the differences between German and U.S. trade fair exhibitors with respect to their knowledge and interaction processes during these events. Other recent research does, however, provide compelling empirical evidence regarding the importance of trade fairs compared to other venues or potential sources of information in firms' innovation processes (Bathelt, 2017).

Table 6.7 Use of trade fairs to search for potential business partners, by nationality of the firm

	U.S. % (n)	Germany % (n)
Yes/sometimes	90% (94)	85% (50)
No	10% (10)	15% (9)
Total	100% (104)	100% (59)

countries on this question. At first glance, this may seem somewhat at odds with the core assumptions of the VoC approach. LMEs like the U.S. tend to rely more on the free movement of technical personnel than on alliances with other firms in order to achieve, for instance, technology transfers (Hall & Soskice, 2001). CMEs like Germany, on the other hand, generally depend more heavily on durable inter-firm alliances in order to reach business and technology-related goals (Hall & Soskice, 2001). As such, more variation between the two countries might have been expected, with German firms placing greater weight than their U.S. counterparts on seeking partners during the trade fair. Upon further consideration, however, it seems plausible that the main difference resides not in whether firms from the two countries look for partners at trade fairs but, rather, in the types of partners they look for at these events. U.S. firms, for example, may be more focused on finding partners in areas such as sales, whereas German firms may be more interested in finding partners in research.

For present purposes, the main insight from Table 6.7 is that the majority of firms from both countries do, in fact, use trade fairs to look for potential partners. This provides sufficient grounds upon which to explore the relationships between the core competencies of firms and those of the partners they seek at trade fairs, as well as how these relationships differ across the two countries. The next sub-section begins by investigating the core capabilities of exhibiting firms from the U.S. and Germany.

6.4.1 Core Capabilities of Firms: General Findings

This section explores the extent to which the core competencies of U.S. and German firms differ. It does so by focusing on seven different focus areas that may be viewed as more or less central to the operations of the firm. These are: (i) innovation focus, (ii) low-cost focus, (iii) top-of-the-line production, (iv) strong research and development capabilities, (v) marketing excellence, (vi) customized production, and (vii) constantly changing product line. Based on the VoC approach, a number of different patterns might be expected. With respect to innovation, for example, the literature suggests that the institutional configurations of LMEs tend to produce strengths in radical innovation, while CMEs tend to enjoy competitive advantages in incremental forms of innovation (Hall & Soskice, 2001; Hancké, 2009). The questionnaire used in this study does not, however, distinguish between different types of innovation. As such, the importance placed on innovation focus is expected to be high from firms in both countries.

With respect to other areas of production, such as customized, low-cost, and top-of-the-line, clear differences between the two countries are expected. German firms, for instance, are expected to provide significantly higher ratings for customized and top-of-the-line production due to the country's specific institutional arrangements, which are well-suited to incremental innovation in high-quality or top-of-the-line manufacturing and customized production (Deeg, 1999). U.S. firms, on the other hand, are expected to put greater emphasis on areas such as low-cost production and constantly changing product line. This is due to the greater labour market flexibility, including less involvement from unions, typically exhibited by LMEs, as well as a less specialized and more mobile workforce (Hall & Soskice, 2001; Ruhs, 2013).

T-tests were used to identify and compare the differences between German and U.S. firms with respect to their own self-identified core capabilities as well as the capabilities they looked for in potential partners (the latter of which will be discussed in the next sub-section). As for the question regarding the firm's own core competencies, firms were asked to rate the importance of the above seven competence areas based on the same 7-point Likert scale that has already been used in the study. Table 6.8 presents the average evaluations of firm and partner capabilities for German and U.S. firms. I found, however, that U.S. firms provided higher ratings on all questionnaire items compared to German firms; it was, thus, important to develop a measure of the relative evaluations (RE) of firms regarding core competencies. This measure is the difference between the country's average evaluation of an individual item (e.g., capacities in customized production) and the overall average evaluation of that respective country across all items. When interpreting the relative evaluations of firms, it is not only the significant differences that are noteworthy, but also those items where the evaluations of Germany and the U.S. are very similar. Moreover, when German firms rate an item higher than their U.S. counterparts, this indicates that these are extremely strong cases. The third column in Table 6.8 below highlights the difference between the relative evaluations of U.S. and German firms.

In rating their own core capabilities, Table 6.8 demonstrates that firms from both Germany and the U.S. had fairly strong core competencies in most of the identified areas, with particular strengths identified in the area of "innovation focus". In terms of both average and relative evaluations, U.S. firms rated "innovation focus" as their dominant core capability (M = 5.89, RE = 0.73), whereas German firms indicated highly important core competencies in both "customized production" (M = 5.76, RE = 0.83) and "innovation focus" (M = 5.70, RE = 0.77). This finding is consistent with the VoC perspective. While it was expected that firms from both capitalist varieties would indicate innovation focus as a highly important core capability, it was in the nature of this innovation focus that I expected to see differences between German and U.S. firms – that is, radical versus incremental innovation.

As previously discussed, radical forms of innovation – meaning significant modifications in product lines, the creation of brand-new products, or dramatic shifts to the production process itself – tend to be better supported by the institutions of LMEs, like the U.S. (Hall & Soskice, 2001; Hancké & Hermmann, 2007; Hancké,

Table 6.8 Comparison of average and relative evaluations of firm and partner capabilities, by country and focus area

	Average evaluations		Relative evaluations		
Firm capabilities	U.S.	Germany	U.S.	Germany	Δ
Innovation focus	5.89	5.70	0.73	0.77	−0.04
Low-cost focus	4.33	3.36	−0.83	−1.57	0.74***
Top-of-the-line production	5.42	5.24	0.26	0.31	−0.05
Strong R&D capabilities	5.49	5.24	0.33	0.31	0.02
Marketing excellence	5.48	5.33	0.32	0.40	−0.08
Customized production	4.88	5.76	−0.28	0.83	−1.11***
Constantly changing product line	4.61	3.86	−0.55	−1.07	0.52*
Overall average evaluation	5.16	4.93	0	0	
Partner capabilities					
Innovation focus	5.35	5.26	0.53	0.65	−0.12
Low-cost focus	4.37	3.50	−0.45	−1.11	0.66**
Top-of-the-line production	5.07	4.77	0.25	0.16	0.09
Strong R&D capabilities	4.84	4.83	0.02	0.22	−0.20
Marketing excellence	4.91	5.04	0.09	0.43	−0.34
Customized production	4.59	5.00	−0.23	0.39	−0.62**
Constantly changing product line	4.58	3.90	−0.24	−0.71	0.47
Overall average evaluation	4.82	4.61	0	0	
N	103	60			

Notes: The cell entries are mean scores, with the third column providing the difference in means between the two countries. The capability variables are based on a 7-point Likert scale, where 1 = "least important" and 7 = "most important"
$^{*}p < 0.10$ $^{**}p < 0.05$ $^{***}p < 0.01$

2009). This has to do with the combination of fluid labour markets, extensive equity markets, and market-based inter-firm relations. As Hall and Soskice (2001, 40) point out, "[f]luid labor markets and short job tenures make it rational for employees to concentrate more heavily on their career than the firm's success and on the development of general skills rather than the industry- or company-specific skills conducive to incremental innovation." With relationships between firms being largely based on markets, it tends to be easier for firms in LMEs to acquire new technologies quickly by, for instance, purchasing other firms, poaching their employees, and licensing new products (Hall & Soskice, 2001, 41).

In contrast to the above description of radical innovation, incremental forms of innovation, which involve ongoing, yet more subtle and gradual improvements to existing product lines and production processes, are generally better supported by the institutional arrangements associated with CMEs, such as Germany (Hall & Soskice, 2001; Hancké & Herrmann, 2007; Hancké, 2009). As Hall and Soskice (2001, 39) explain, this has to do with the "relational requirements of company endeavors". Incremental innovation is easier to accomplish in contexts where "the workforce [...] is skilled enough to come up with such innovations, secure enough to risk suggesting changes to products or process that might alter their job situation, and endowed with enough work autonomy to see these kinds of improvements as a dimension of their job" (Hall & Soskice, 2001, 39). The institutional arrangements characteristic of CMEs generally provide the necessary support structure for these relational requirements. Therefore, the fact that the ratings of German firms were similar to those of U.S. firms with respect to having core capabilities in innovation focus supports the overall argument of this study.

Beyond the shared emphasis on "innovation focus", there were a number of notable differences between firms from Germany and the U.S. with respect to the following three areas of core competence: (i) "low-cost focus", (ii) "constantly changing product line", and (iii) "customized production". Table 6.8 summarizes these differences. In the area of low-cost focus, for example, U.S. firms were found to be significantly more likely than German firms to suggest strong core competencies. The results on this front are highly significant at the 1% level, with nearly a full point difference on the 7-point Likert scale. Similarly, U.S. firms reported significantly higher core competencies than German firms in the area of constantly changing product line. Here, the differences are also significant, but at a more modest 10% level. Based on the VoC perspective, the differences between the two countries in low-cost focus and constantly changing product line were expected. These differences relate to the embeddedness of U.S. firms in a market-oriented system that promotes rapid reconfigurations of existing technologies or the adoption of new technologies through short-term labour contracts, a highly mobile workforce, minimal involvement of workers in wage and other related negotiations, as well as equity markets that impose only limited restrictions on firm mergers or acquisitions (Hall & Soskice, 2001). Taken together, these conditions provide strong incentives for firms to deliver rapid innovations and constantly changing product lines. As will be discussed in greater detail below, the greater flexibility of the labour market in LMEs, including less involvement from unions and a less specialized, more mobile workforce supplies the necessary support structure to maintain low-cost production systems (Hall & Soskice, 2001; Ruhs, 2013).

As Table 6.8 indicates, "customized production" represents another highly significant area of difference between the two countries, with German firms (M = 5.76, RE = 0.83) indicating much higher core competencies in this area than their U.S. counterparts (M = 4.88, RE = −0.28). When focusing on the difference between the relative evaluation scores, the results are quite striking. The finding is significant at the 1% level. This result reflects the assumptions of the VoC model in that CMEs, such as Germany, tend to foster deep investments in the vocational training system, a highly specialized workforce that often includes lengthy apprenticeships in specific industries, as well as long-term commitments and collaborative arrangements between firms, their employees, owners, and suppliers – all of which serve to support incremental innovations in customized production (Deeg, 1999; Hall & Soskice, 2001;

Boschma & Capone, 2015). The legacy of Germany's *Handwerk*[4] system is a powerful example of continued specialization in customized production despite ongoing pressures towards liberalization and deregulation (Streeck, 1989; Thelen, 2004).

In addition to the core capabilities discussed above, there were also three areas of competence that did not reveal significant differences between the two countries – "top-of-the-line production", "strong R&D capabilities", and "marketing excellence". While not significant, the findings in these areas generally support the VoC thesis when one focuses attention on the relative evaluations of firms. For example, German firms rated capabilities in top-of-the-line production slightly higher than their U.S. counterparts, as was expected for similar reasons as in customized production. In terms of marketing excellence, the RE measure shows a slightly stronger rating from German than U.S. firms, which is somewhat surprising given the nature and centrality of market forces in LMEs and the relationship between marketing excellence and competition in such environments. Yet, the nearly identical ratings on strong R&D capabilities by firms from both countries generally support expectations.

6.4.2 Core Capabilities of Partners: General Findings

Having discussed the striking and significant differences between firms from the U.S. and Germany with respect to their own core capabilities, it is now possible to explore the hypothesis that, in the context of the trade fair, firms from both LMEs and CMEs will generally seek partners with core capabilities that closely match their own areas of competence. This is based on the argument that actors will internalize the system logics associated with the specific institutional frameworks of the country in which they are embedded and engage in behaviours that largely reflect these system logics. Table 6.8 above summarizes the findings. In terms of the core capabilities of potential business partners, the data revealed three important areas of difference between the two countries: "low-cost focus", "customized production", and "constantly changing product line". The first two are significant at the 5% level, while the last core capability – constantly changing product line – shows a clear difference with respect to the relative evaluations of firms from the two countries.

When it came to seeking out partners during the trade fair, U.S. firms placed significantly greater emphasis than did German firms on finding partners that have

[4]For Thelen (2004, f.n.1), the critical distinction is between "industry" and *Handwerk:* "The difference between the *Handwerk* (handicraft) sector and industry was traditionally partly a matter of size (handicraft firms typically being smaller) but was linked also to the nature of production". In the *Handwerk* sector, the nature of production historically involved skilled manual labour that contained little mechanization (Kocka, 1984; Thelen, 2004). Providing a different interpretation of *Handwerk* as a social and legal institution, Streeck (1989, 62) suggests that "most of what is called 'small business' in other countries falls in Germany in the legal category of *Handwerk.* The least awkward equivalent of *Handwerker* in English is probably 'artisan'; a close competitor might be 'independent craftsman' [...]".

specific core competencies in low-cost focus (U.S. firms: M = 4.37, RE = −0.45 vs. German firms: M = 3.50, RE = −1.11). On the other hand, German firms indicated a significantly higher interest in partners with core competencies in customized production (U.S. firms: M = 4.59, RE = −0.23 vs. German firms: M = 5.00, RE = 0.39). These findings are consistent with the VoC thesis. Low-cost production systems, for example, are a common and expected feature of market-oriented environments that are characterized by a high degree of competition, considerable labour market flexibility, and highly mobile capital (Hall & Soskice, 2001; Guider, 2007; Freeman & Mirilovic, 2016). Recent studies in immigration policy provide an interesting example of how governments and policymakers in LMEs can moderate wage growth and maintain relatively low-cost production systems through the employment of flexible migrant workers (Ruhs, 2016). The recent work of Ruhs (2016, 448) suggests that while, in CMEs, "there are likely to be strong pressures, partly through the involvement of unions in the design and implementation of labor immigration policies, to employ migrants at the collectively agreed-on wage", such pressures are largely absent in the context of LMEs.

With respect to customized production, the fact that German firms were significantly more likely than U.S. firms to look for partners with core capabilities in this area is also consistent with the VoC literature, as well as the study's empirical findings thus far. As noted above, the institutional configurations of CMEs tend to encourage a highly specialized workforce, long-term commitments, and a high amount of collaboration between firms and their employees, owners, and suppliers. Such conditions support ongoing improvements and incremental innovations in customized production (Deeg, 1999; Hall & Soskice, 2001; Boschma & Capone, 2015).

The finding regarding constantly changing product line also fits with my expectations. As was previously mentioned, the institutional arrangements associated with LMEs tend to support rapid innovations and constantly changing product lines, whereas those of CMEs support incremental innovations and ongoing improvements to ensure that the highest quality of an established product line is maintained (Hall & Soskice, 2001). Within the context of LMEs, competitive and weakly regulated product markets demand "a capacity for taking risks on new product strategies and for the rapid implementation of such strategies [...]" (Hall & Soskice, 2001, 39). It was therefore not surprising that firms from the U.S. would place more weight on constantly changing product line than did their German counterparts.

6.4.3 Relationship Between Firm and Partner Capabilities: General Findings

Having identified clear and significant cross-national differences in the core capabilities of firms as well as the competencies they look for in potential partners at the trade fair, an important next step is to focus more specifically on the relationship between these two sets of competencies – that of the firm and that of its partners.

Table 6.9 below provides a simple correlation of firms' average evaluations of their own capabilities by those of potential partner capabilities. This is done for each of the seven focus areas for both Germany and the U.S. These correlations provide an initial sense of the relationship between a firm's core strengths and those sought in potential business partners during the trade fair. Based on the VoC approach, the expectation here was that firms at trade fairs would be primarily guided by their pre-existing production contexts and knowledge bases and, as such, would seek partners that closely reflected their own core competencies.

Through a series of Pearson correlation tests between firm and partner capacities (based on 1–7 Likert scales measuring importance in competency), Table 6.9 illustrates that, in most cases, there was a clear tendency for firms from both countries to look for partners with core capabilities that closely matched or reinforced their own. Nevertheless, the evidence of this relationship was somewhat stronger for German firms than U.S. firms. With the exception of innovation ($r_{us} = 0.46$, $r_{germany} = 0.41$) and low-cost focus ($r_{us} = 0.51$, $r_{germany} = 0.50$), for example, German firms exhibited a clearer tendency than their U.S. counterparts to look for the same or very similar capabilities in potential partners at the trade fair. In addition to displaying less obvious ties between their own core competencies and those they looked for in partners, the table below suggests that in a couple of areas – namely, top-of-the-line production ($r_{us} = 0.22$, $p < 0.06$) and R&D capabilities ($r_{us} = 0.24$, $p < 0.03$) – the correlation is rather modest for U.S. firms. Indeed, when I looked further into the distribution of responses and mean scores in these competence areas, I found that U.S. firms placed greater importance on themselves having competencies in top-of-the line production and R&D capabilities than they did on finding partners with similar strengths. When comparing the two countries, German firms, thus, displayed clearer areas of convergence between their own and their partners' capabilities than did U.S. firms. This was most pronounced in the following areas: top-of-the-line production ($r_{germany} = 0.47$, $p < 0.001$), R&D capabilities ($r_{germany} = 0.43$, $p < 0.01$), and constantly changing product line ($r_{germany} = 0.68$, $p < 0.001$). With U.S. firms, the

Table 6.9 Correlation coefficients of firm capabilities with partner capabilities, by country and focus area

	United States	Germany
Innovation focus	0.463^{***}	0.413^{***}
Low-cost focus	0.506^{***}	0.503^{***}
Top-of-the-line production	0.218	0.474^{***}
Strong R&D capabilities	0.239^{**}	0.434^{***}
Marketing excellence	0.260^{**}	0.343^{**}
Customized production	0.473^{***}	0.521^{***}
Constantly changing product line	0.510^{***}	0.684^{***}
N	103	60

Notes: cell entries are Pearson correlation coefficients and are based on a 7-point Likert scale, where 1 = least important and 7 = most important

$^{*}p < 0.10$ $^{**}p < 0.05$ $^{***}p < 0.01$

strength of the convergence correlation was only strong in one of these three areas – constantly changing product line ($r_{us} = 0.51$, $p < 0.001$). And, still, the correlation was notably stronger in the German case.

These findings suggest that U.S. firms tend to be associated with more general competencies that allow them to shift their production. It might also be argued that these firms look for broader competencies in their business partners because they are less specialized and more interested in developing generic and switchable assets. These results are broadly in keeping with the VoC literature and, specifically, the argument that the institutional setting of LMEs creates incentives for firms to focus less on developing specializations and more on developing "assets whose value can be realized if diverted to other purposes" (Hall & Soskice, 2001, 17). For present purposes, however, what is most critical to stress is that the core competencies of firms and the partners they seek at trade fairs tend to be very closely matched.

6.4.4 Findings Related to Firm Age, Size, and Sector

The data presented in Table 6.8 provided a general sense of the types of core capabilities German and U.S. firms possessed as well as the kinds of competencies they looked for in potential partner firms. Table 6.9 built further on these findings to reveal that these two sets of core competencies usually mirrored each other. Next, in an effort to engage more deeply with the findings on core competencies, I revisited the relationships that were previously shown to be significant in Table 6.8, but this time controlled for additional firm characteristics – specifically, the firm's age, size, and sector. T-tests were used to identify and compare differences in the average evaluations of firms. Table 6.10 presents these findings.

As displayed in the table, sector stands out as a significant area of difference. Compared to firms in the machinery sector, electronics firms were more likely to give a high rating to low-cost production and constantly changing product line as core competencies for both the firm itself and its potential partners. These findings are significant at the 5% and 1% levels. In contrast, firms in the machinery sector rated customized production much more highly than their counterparts in the electronics sector, at a significance level of 5%. Firm age and size did not demonstrate a clear relationship to either the self-identified core competencies of the firm or those it seeks out in potential business partners.

The first result regarding sector can be partly explained by the fact that electronics is a fast-moving technology sector that generally relies more on rapid than incremental processes of innovation (Akkermans et al., 2009; Boschma & Capone, 2015). As discussed more extensively above, LMEs, such as the U.S., are better at supporting rapid innovation than are CMEs like Germany (Hall & Soskice, 2001). It follows that the U.S. would enjoy comparative advantages in this sector and that low-cost production and constantly changing product line would be part of this broader constellation of institutional complementarities associated with LMEs.

Table 6.10 Average evaluations of firm and partner capabilities, by firm age, size, sector, and focus area

	Age			Size			Sector		
Focus area	[a] Post-globalization	[b] Pre-globalization	Δ	[c] Small	[d] ML	Δ	Electronics	Machinery	Δ
Firm capabilities									
Innovation focus	5.97	5.72	0.27	5.70	5.96	−0.26	5.86	5.75	0.11
Low-cost focus	4.07	3.95	0.12	4.05	3.84	0.21	4.18	3.61	0.57**
Customized production	5.10	5.25	−0.15	5.02	5.41	−0.39	4.99	5.58	−0.59**
Changing product line	4.30	4.35	−0.05	4.39	4.38	0.01	4.65	3.78	0.87***
Partner capabilities									
Innovation focus	5.42	5.24	0.18	5.33	5.31	0.02	5.37	5.24	0.13
Low-cost focus	4.16	3.99	0.17	4.00	4.07	−0.07	4.36	3.57	0.79***
Customized production	4.67	4.74	−0.07	4.57	4.91	−0.34	4.75	4.74	0.01
Changing product line	4.43	4.20	0.23	4.50	3.91	0.59*	4.69	3.69	1.00***
N	94	60		93	51		60	104	

Notes: T-tests were used to identify and compare differences in the average evaluations of firms. These evaluations are based on a 7-point Likert scale, where 1 = "least important" and 7 = "most important"

*$p < 0.10$ **$p < 0.05$ ***$p < 0.01$

[a]Post-Globalization refers to firms that were established in 1990 or later, corresponding with the start of the globalization period

[b]Pre-Globalization refers to firms that were established after 1990

[c]Small firms are those that have 100 or fewer employees

[d]ML denotes Medium-sized and Large firms. In this study, ML firms are those that have more than 100 employees

6.4.5 Results of Additional Control Variables

In addition to differentiating by firm age, size, and sector, this section follows the logic of the previous bivariate analysis on information sources (Sect. 6.2) in distinguishing among further firm characteristics and behaviours that might influence the core competencies of firms and their potential partners. To remind the reader, these additional firm characteristics are: (i) the international versus domestic focus of the firm's operations; (ii) whether or not the firm engages in R&D collaborations with other firms; (iii) whether or not the firm has merged with or acquired another firm in the past 3 years; (iv) whether or not the firm co-operates with external organizations in industry standard-setting; and (v) whether or not the firm has received government support to participate in the trade fair.

While Table 6.11 identifies a number of different relationships, the discussion below focuses on some of the more significant ones. First, firms that engage in R&D collaborations with other organizations tended to place higher importance on innovation focus than firms that do not engage in such collaborations. This is true with respect to both the core competencies of the firms and the competencies they look for in partners, with significance levels at the 1% and 5% levels respectively. This is perhaps not surprising given the intersections between R&D activities and innovation and economic growth (Fagerberg, 1987; Lichtenberg, 1992; Ulku, 2004). Yet, the relationship between R&D and innovation is not straightforward, partly because R&D activities can have different functions and, as already discussed, innovation can mean different things in different contexts. Incremental innovation in producer-user contexts, for instance, may not rely on R&D. Moreover, while considerable attention has been paid to the relationship between R&D activities and innovation performance (e.g., Fagerberg, 2005), here the focus is slightly different in that I am specifically interested in inter-firm R&D collaborations and their relationship to a firm's core capabilities in innovation.

A similar pattern exists for firms that co-operated with external organizations in industry standard-setting. These firms placed significantly higher importance on innovation than firms that were not involved in industry standard-setting. Again, this is true with respect to both the core competencies of the firm and the competencies firms look for in potential partners, with significance levels of 1% and 5%, respectively. In addition, firms that participated in industry standard-setting viewed customized production as more important than firms that did not participate in such standard-setting practices. But this pattern is only apparent with regards to the core competencies of the firm itself, not with respect to the competencies of potential partners. Finally, firms with mainly domestic-oriented operations as well as firms that had not received government support for trade fair participation were more likely than their counterparts to place a high level of importance on constantly changing product line. There was no clear relationship between firms that merged with or acquired another firm in the past 3 years and the core competencies of either firms or partners.

Table 6.11 Average evaluations of firm and partner capabilities, by specific firm characteristics and behaviours and focus area

	[a] Internationalization			[b] R&D collaboration			[c] Merged or acquired			[d] Industry standard-setting			[e] Received government support		
	International	Domestic	Δ	No	Yes	Δ	No	Yes	Δ	No Participation	Yes Participates	Δ	No	Yes	Δ
Firm capabilities															
Innovation	5.89	5.78	0.11	5.47	6.14	−0.67***	5.83	5.78	0.05	5.51	6.12	−0.61***	5.78	5.88	−0.10
Low-cost	3.95	4.00	−0.05	3.71	4.20	−0.48*	3.96	4.02	−0.06	4.08	3.84	0.22	4.08	3.64	0.44
Customized production	5.29	5.17	0.12	5.04	5.33	−0.29	5.16	5.29	−0.13	4.95	5.51	−0.56**	5.14	5.36	−0.22
Changing product	4.16	4.54	−0.38	4.29	4.43	−0.14	4.35	4.29	0.06	4.17	4.49	−0.32	4.54	3.84	0.70*
Partner capabilities															
Innovation	5.26	5.38	−0.12	5.07	5.57	−0.50**	5.26	5.46	−0.20	5.02	5.57	−0.55**	5.40	4.82	0.58*
Low-cost	3.90	4.24	−0.34	3.88	4.29	−0.41	4.03	4.19	−0.16	4.03	4.07	−0.04	4.13	3.80	0.33
Customized production	4.65	4.85	−0.20	4.60	4.93	−0.33	4.76	4.68	0.08	4.57	4.89	−0.32	4.85	4.06	0.79*
Changing product	4.00	4.66	−0.66**	4.18	4.48	−0.30	4.30	4.36	−0.06	4.17	4.41	−0.24	4.50	3.58	0.92**
N	82	80		73	87		114	49		76	83		125	25	

Notes: T-tests were used to identify and compare differences in the average evaluations of firms. These evaluations are based on a 7-point Likert scale, where 1 = "least important" and 7 = "most important"

$^{*}p < 0.10$ $^{**}p < 0.05$ $^{***}p < 0.01$

[a]Internationalization refers to whether the operations of the firm are mainly international or domestic

[b]This refers to whether the firm engages in R&D collaborations with other firms

[c]Has the firm merged with or acquired another firm in the past 3 years?

[d]Refers to whether the firm co-operates with external organizations in industry standard setting

[e]Has the firm ever used government support to participate in trade fairs?

6.4.6 Regression Results

Although several factors seem to be involved, the discussion thus far still suggests that the nationality of the firm plays a decisive role in both the core competencies of firms as well as the competencies firms look for in potential partners. In order to test whether the observed country differences remain when the independent effects of the other relevant factors are taken into account, multivariate linear regression analyses were conducted. Table 6.12 below includes as dependent variables the three core competencies that showed significant differences between the U.S. and Germany in the initial comparison of means test (see Table 6.8). These are: (i) "low-cost focus"; (ii) "customized production"; and (iii) "constantly changing product line". Given the study's related interest in innovation processes, I also included as a fourth dependent variable the core capability of "innovation focus". As in the previous regression analysis, the independent variables are the specific characteristics or behaviours of firms explored above in Tables 6.9 and 6.11. A variable on the survey distribution mode was also added in order to explore any possible biases associated with the Internet versus paper survey formats.

The results of the multivariate linear regression, summarized in Table 6.12, reveal some valuable insights. Most importantly, for the present analysis, is that even after considering the independent effects of other relevant variables, U.S. and German firms differed with regards to the core capabilities they possessed. More specifically, firms from the U.S. were significantly more likely than their German counterparts to have core competencies in the area of low-cost focus, whereas German firms were significantly more likely than U.S. firms to have core competencies in the area of customized production. These findings are consistent with the assumptions of the VoC approach regarding the specific sets of institutional supports underlying national production strategies.

In the context of LMEs, for instance, where economic activity is largely coordinated through corporate independence, the competitiveness of firms tends to be built on both low-cost production and rapid technological innovation (Hall & Soskice, 2001; Guider, 2007). Yet, in CMEs like Germany, economic activity is largely coordinated through non-market mechanisms (e.g., labour unions) and, thus, the competitiveness of firms tends to be based on "incremental innovation in high quality manufacturing and customized production" (Deeg, 1999, 6; see also, Hall & Soskice, 2001; Guider, 2007). It is, therefore, the specific institutional arrangements associated with LMEs that can help explain the greater emphasis U.S. firms place on low-cost production as a core capability. In the case of Germany, it is the national institutional framework again that serves to enable the long-term commitments and cooperative arrangements among the firm, its employees, owners, and suppliers that are needed for customized production (Deeg 1999, 6).

Beyond the striking results regarding cross-national differences in the core capabilities of firms, Table 6.12 also sheds light on the factors associated with innovation focus. Here, as at the bivariate level, it is not country but engagement in R&D collaborations that seems to be most important. More specifically, firms that engaged

Table 6.12 Regression analysis: factors associated with the core capabilities of firms

	Innovation focus	Low-cost focus	Customized production	Changing product line
Country				
Germany vs. U.S.	0.24 (0.28)	1.20 (0.42)***	−0.74 (0.42)*	0.33 (0.43)
Firm characteristics				
Trade fair type				
Machinery vs. electronics	−0.20 (0.29)	−0.16 (0.44)	−0.74 (0.44)*	0.44 (0.45)
Size of firm				
Small vs. medium/large	0.10 (0.22)	0.01 (0.33)	0.09 (0.33)	0.03 (0.34)
Age of firm				
Pre-1990 vs. 1990 or later	0.15 (0.23)	−0.13 (0.34)	0.53 (0.34)	−0.13 (0.35)
Firm behaviour				
International vs. domestic	−0.27 (0.23)	−0.54 (0.34)	0.29 (0.34)	−0.09 (0.35)
Engaged in R&D	0.64 (0.22)***	0.51 (0.33)	0.23 (0.33)	−0.24 (0.34)
Merged in past 3 years	−0.21 (0.22)	−0.13 (0.33)	0.00 (0.33)	−0.13 (0.34)
Participated in industry standard-setting	−0.38 (0.22)*	−0.55 (0.33)*	0.72 (0.33)**	0.50 (0.34)
Received government support	0.06 (0.27)	−0.33 (0.41)	0.11 (0.46)	−0.34 (0.42)
Survey distribution mode				
Internet vs. paper	−0.44 (0.22)	0.10 (0.33)	−0.32 (0.33)	−0.00 (0.34)
R-squared	0.20	0.12	0.15	0.06
N	127	128	127	128

Unstandardized coefficients presented in cells with standard errors in parentheses
Dependent Variables coded on a 7-point scale, with 1 = "least important" and 7 = "most important"
The independent and control variables are coded as follows: U.S. = '1', Germany = '0'; Machinery = '0', Electronics = '1'; Small firm = '0', ML firm = '1'; Pre-globalization firm = '0', post-globalization firm = '1'; International focus of operations = '0', Domestic focus = '1'; "yes" engaged in R&D = '0', "no" did not engage in R&D = '1'; "yes" merged with/acquired another firm in the last 3 years = '0', "no" did not merge with another firm in the last 3 years = '1'; "yes" participates in industry standard-setting '0', 'no" does not participate in industry standard-setting '1'; "yes" received government support to participate in the trade fair = '0', "no" did not receive government support = '1'; Internet survey distribution = '0', Paper, F2F distribution = '1'
$^{*}p < 0.10$ $^{**}p < 0.05$ $^{***}p < 0.01$

in inter-firm R&D collaborations were significantly more likely than firms that did not engage in such R&D collaborations to view innovation focus as an important core capability. This finding makes sense on an intuitive level, at least, given the multiple advantages associated with engaging in inter-firm R&D collaborations, including quicker access to creative new combinations of knowledge (Grant & Baden-Fuller, 2004; Linnarson, 2005), increased access to technological and learning opportunities, as well as opportunities for accelerating product development (Gilsing et al., 2007; Quintana-García & Benavides-Velasco, 2011).

Beyond this, after taking into account all the variables, none of them were particularly helpful in understanding the firms' core competencies with respect to continuously changing product line. This is exemplified by the low R-squared score, suggesting that there are likely other factors involved that this study was not able to capture. Finally, as in the multivariate regression analysis conducted for information sources (see Table 6.4), survey distribution mode did not have a significant impact on the core competencies of firms.

Having explored the independent effects of various factors on the core competencies of the firms themselves, a next important step is to conduct a multivariate linear regression analysis of the core competencies firms look for in potential partners at the trade fair. Using the same dependent and independent variables as described above, the results of the regression analysis are summarized below in Table 6.13.

First, and most importantly for the present study, country remains a clear and significant factor in at least one of the core capability areas – customized production. Consistent with prior regression results presented above, German firms were significantly more likely than their U.S. counterparts to look for partners with specific strengths in the area of customized production. This result fits well with the VoC model, as indicated in the above description regarding the relationship between CMEs and customized production patterns.

Beyond the continued relevance of country, other factors seem to play a role in the core competencies firms look for in potential partners. For example, firms that co-operated with external organizations in industry standard-setting were more likely than those that did not to look for partners with core capabilities in innovation focus. With respect to continuously changing product line, however, none of the factors offered much insight. This is consistent with previous multivariate regression results on the core competencies of firms (see Table 6.12). However, it raises questions about what factors might be better able shed light on this area.

Finally, it is important to note that survey mode did have an impact in two cases – namely, innovation and customized production. Those that filled out the paper questionnaire through F2F distribution were more likely to look for partners with core capabilities in these two areas than were those that completed the questionnaire online. This finding is puzzling given that survey mode did not play a role in either of the previous regression analyses. As noted elsewhere, contextual factors associated with filling out the questionnaire while in the midst of the trade fair compared to completing the survey alone in front of one's computer might have something to do with it. Yet, to the extent that such an argument is true, it is unclear why the impact of survey mode would be restricted to innovation focus and customized production to the exclusion of low-cost production and constantly changing product line. Despite this anomaly in the findings, the overarching hypothesis seems to hold: Country remains a strong factor in determining differences in the core competencies of both firms and their partners at international trade fairs.

Table 6.13 Regression analysis: factors associated with the core capabilities of potential partners

	Innovation focus	Low-cost focus	Customized production	Changing product line
Country				
Germany vs. U.S.	−0.17 (0.39)	0.55 (0.44)	−0.89 (0.42)**	−0.13 (0.46)
Firm characteristics				
Trade fair type				
Machinery vs. electronics	0.26 (0.41)	0.19 (0.46)	0.49 (0.44)	0.48 (0.50)
Size of firm				
Small vs. medium/large	−0.19 (0.29)	0.16 (0.34)	0.44 (0.31)	−0.44 (0.34)
Age of firm				
Pre-1990 vs. 1990 or later	−0.09 (0.31)	0.07 (0.36)	−0.14 (0.34)	0.11 (0.37)
Firm behaviour				
International vs. domestic	−0.07 (0.30)	0.04 (0.35)	0.21 (0.33)	0.37 (0.36)
Engaged in R&D	0.34 (0.30)	0.18 (0.35)	0.15 (0.34)	−0.04 (0.36)
Merged in past 3 years	−0.02 (0.30)	0.06 (0.35)	−0.06 (0.31)	0.15 (0.35)
Participated in industry standard-setting	0.53 (0.29)*	−0.04 (0.34)	0.57 (0.33)*	0.35 (0.35)
Received government support	−0.44 (0.35)	−0.10 (0.46)	−1.03 (0.43)**	−0.49 (0.45)
Survey distribution mode				
Internet vs. paper	0.78 (0.33)**	0.33 (0.37)	0.88 (0.35)**	0.17 (0.37)
R-squared	0.16	0.06	0.19	0.10
N	102	108	103	106

Unstandardized coefficients presented in cells with standard errors in parentheses
Dependent Variables coded on a 7-point scale, with 1 = "least important" and 7 = "most important"
The independent and control variables are coded as follows: U.S. = '1', Germany = '0'; Machinery = '0', Electronics = '1'; Small firm = '0', ML firm = '1'; Pre-globalization firm = '0', post-globalization firm = '1'; International focus of operations = '0', Domestic focus = '1'; "yes" engaged in R&D = '0', "no" did not engage in R&D = '1'; "yes" merged with/acquired another firm in the last 3 years = '0', "no" did not merge with another firm in the last 3 years = '1'; "yes" participates in industry standard-setting '0', 'no" does not participate in industry standard-setting '1'; "yes" received government support to participate in the trade fair = '0', "no" did not receive government support = '1'; Internet survey distribution = '0', Paper, F2F distribution = '1'
*$p < 0.10$ **$p < 0.05$ ***$p < 0.01$

6.5 Measuring Inter-firm Homogeneity Versus Complementarity

The above discussion sheds important light on the nature of firm and partner core capabilities. Yet, the actual *degree* of difference between the core capabilities a firm possesses and those it looks for in potential partners still requires further explication. In an effort to develop a more precise measure of inter-firm homogeneity versus heterogeneity or complementarity, I created a *difference variable* on a scale of

0–6, where complete homogeneity (in terms of the firm's self-identified competencies versus those that it searches for in partners) equals 0, and complete heterogeneity or complementarity (in the same firm and partner core competencies) equals 6. This was accomplished by subtracting one evaluation score (i.e., for the firm) from the other (i.e., for the partners). The resulting homogeneity/heterogeneity scale is, thus, essentially the difference between the two variables, either in the one direction (i.e., towards homogeneity) or the other direction (i.e., towards heterogeneity or complementarity).

Tables 6.14, 6.15, 6.16, 6.17 and 6.18 investigate the degree of homogeneity vs. complementarity between exhibiting firms and their potential partners across countries in each of the core capability areas that were revealed as significant in the previous analysis. These areas are: low-cost focus, customized production, and constantly changing product line. In keeping with the discussion up until this point, innovation focus has also been included due to its importance in processes of learning, innovation, and knowledge exchange between firms. Finally, I also included strong R&D capabilities at the end of the discussion because it emerged as the most notable example of cross-national difference on the homogeneity/complementarity measure.

6.5.1 *Innovation Focus*

Starting with innovation focus, Table 6.14 shows that the majority of firms in both the U.S. and Germany demonstrated a high to moderate degree of homogeneity between the core competencies of the firm and those of its potential partners. In other words, if a firm rated its own innovation focus capabilities as a 6 on the 1–7 Likert scale, it tended to rate the same capability equally or about as high for the partners it looked for at trade fairs. Looking at the measures for complete and moderate homogeneity (0 and 1), for instance, the data suggest that there was hardly any difference (10% points) in the number of German firms over U.S. firms that reported this level of homogeneity. Similarly, in the most heterogeneous category, which

Table 6.14 Difference between a firm's central capabilities and the capabilities it seeks in potential partners: *Innovation Focus*

	USA # (%)	Germany # (%)
Complete homogeneity	37 (47)	19 (38)
Moderately homogeneous	21 (27)	23 (46)
Heterogeneous	20 (26)	8 (16)
Chi-square	5.14*	
N	128	

Notes: 0 = Complete Homogeneity; 1 = Moderately Homogeneous; and 2–6 = Heterogeneous
$^{*}p < 0.10$ $^{**}p < 0.05$ $^{***}p < 0.01$

includes responses ranging from 2 through 6 (with 6 being complete heterogeneity), Table 6.14 shows that close to the same number of German and U.S. firms demonstrated this degree of inter-firm heterogeneity. These results are modestly significant at the 10% level. A larger number of U.S. firms than German firms clustered in the heterogeneous category with respect to innovation focus, which is consistent with the view that LMEs tend to encourage firms that are more diversified and able to adjust more readily to competitive pressures and changes in the market (Boschma & Capone, 2015). However, the overarching trend for both German and U.S. firms is one of homogeneity between firm and partner capabilities in innovation focus.

6.5.2 *Low-Cost Focus*

Table 6.15 describes country differences in inter-firm homogeneity/complementarity in the area of low-cost focus. Recall that previous bivariate and multivariate analyses revealed clear differences between U.S. and German firms with regards to the importance placed on low-cost production, a finding which reflects the different institutional configurations and system logics of their respective home countries. As such, we might have expected U.S. firms to lean more strongly towards homogeneity on this front. Yet, as the table below suggests, a larger number of U.S. firms compared to German firms exhibited heterogeneity in relation to their partner's core competencies. While more than half (55%) of German firms displayed complete homogeneity between firm and partner capabilities on low-cost focus, just over one-third (36%) of U.S. firms demonstrated the same. And, at the other end of the spectrum, nearly one-quarter (24%) of U.S. firms revealed substantial heterogeneity between their own and their partner's capabilities, while only 13% of German firms exhibited a similar degree of heterogeneity. These results are moderately significant at the 10% level.

A general picture is, thus, emerging with regards to the degree of homogeneity or heterogeneity demonstrated by firms from the two countries. Regardless of the area of focus, U.S. firms seem to display slightly greater heterogeneity with respect to the match between their own core competencies and the competencies of their

Table 6.15 Difference between a firm's central capabilities and the capabilities it seeks in potential partners: *Low-cost Focus*

	USA # (%)	Germany # (%)
Complete homogeneity	33 (36)	26 (55)
Moderately homogeneous	36 (40)	15 (32)
Heterogeneous	22 (24)	6 (13)
Chi-square	5.11*	
N	138	

Notes: 0 = Complete Homogeneity; 1 = Moderately Homogeneous; and 2–6 = Heterogeneous
*$p < 0.10$ **$p < 0.05$ ***$p < 0.01$

partners. This means that U.S. firms that generally placed a high value on low-cost focus did not necessarily look for low-cost partners; they also looked for partners with other competencies. German firms, on the other hand, tended to exhibit greater homogeneity on this front. As discussed above in relation to innovation focus, this tendency is broadly consistent with the VoC literature, as the institutional framework supporting economic behaviour in the U.S. generally provides fewer incentives for firms to specialize and stronger incentives to develop switchable assets. The opposite is true for firms operating in the German CME.

6.5.3 Customized Production

Table 6.16 focuses on the area of customized production. Here, the data present a similar story as the past two tables insofar as there is a general tendency towards inter-firm homogeneity in both the U.S. and Germany. In other words, firms with core capabilities in customized production in both countries tended to look for similar core competencies in their partners, at 70% and 73%, respectively, on the homogeneity/heterogeneity scale (complete and moderate homogeneity combined). Yet, as demonstrated in the past bivariate and multivariate analyses, German firms were significantly more focused on this area than were U.S. firms, a finding which fits with the respective institutional settings. Add to this the fact that more than half of German firms in this area looked for complete homogeneity in their partners in terms of their competence structure and the picture of specialization in this competence area becomes even more striking.

6.5.4 Constantly Changing Product Line

Table 6.17 looks at the degree of inter-firm homogeneity/heterogeneity in the area of constantly changing product line. The previous bivariate analysis (see Table 6.8) revealed significant cross-national differences in this area, with U.S. firms placing

Table 6.16 Difference between a firm's capabilities and those it seeks in partners: *Customized Production*

	USA # (%)	Germany # (%)
Complete homogeneity	27 (34)	25 (51)
Moderately homogeneous	29 (36)	11 (22)
Heterogeneous	24 (30)	13 (27)
Chi-square	4.24	
N	129	

Notes: 0 = Complete Homogeneity; 1 = Moderately Homogeneous; and 2–6 = Heterogeneous
$^{*}p < 0.10$ $^{**}p < 0.05$ $^{***}p < 0.01$

Table 6.17 Difference between a firm's capabilities and those it seeks in partners: *Changing Product Line*

	USA # (%)	Germany # (%)
Complete homogeneity	32 (38)	25 (52)
Moderately homogeneous	28 (33)	14 (29)
Heterogeneous	25 (29)	9 (19)
Chi-square	2.99	
N	133	

Notes: 0 = Complete Homogeneity; 1 = Moderately Homogeneous; and 2–6 = Heterogeneous
$^{*}p < 0.10$ $^{**}p < 0.05$ $^{***}p < 0.01$

much more importance on changing product line, for both themselves and potential partners, than their German counterparts. Yet, the significance of this relationship did not hold in the multivariate analyses (Tables 6.12 and 6.13). The general tendency in Table 6.17 mirrors the above description of customized production. The degree of heterogeneity was more apparent with U.S. firms than it was with German firms, again reflecting the institutional contexts in which these firms operate. Yet, in general, firms from the two countries demonstrated a high degree of homogeneity in the sense that firms with strengths in constantly changing product lines tended to choose partners with core competencies that matched their own.

6.5.5 *Strong R&D Capabilities*

Although the previous bivariate and multivariate analyses did not indicate significant cross-national differences in the area of strong R&D capabilities (see Tables 6.8, 6.12, and 6.13), the significance of the findings from the homogeneity/heterogeneity scale below are worth mentioning. While Table 6.18 indicates that the majority of firms from the U.S. and Germany looked for partners with similar competence profiles, 62% and 83% respectively, in the most heterogeneous category, there are 20 percentage points difference between U.S. and German firms. More specifically, heterogeneity between the core competencies of firms and partners in the area of strong R&D focus was more prevalent among U.S. firms (38%) than German firms (17%). These results are significant at the 5% level.

To summarize, the above analysis of inter-firm homogeneity/heterogeneity reveals two strong findings. First, firms from both countries had a tendency to look for partners that possessed core competencies that were similar to their own. This tendency was apparent across all five of the competence domains explored above – innovation focus, low-cost production, customized production, constantly changing product line, and strong R&D capabilities. Second, in some cases, U.S. firms were somewhat less likely than German firms to search for partners with core

Table 6.18 Difference between a firm's capabilities and capabilities it seeks in partners: *R&D Focus*

	USA # (%)	Germany # (%)
Complete homogeneity	29 (33)	23 (48)
Moderately homogeneous	25 (29)	17 (35)
Heterogeneous	33 (38)	8 (17)
Chi-square	6.76**	
N	135	

Notes: 0 = Complete Homogeneity; 1 = Moderately Homogeneous; and 2–6 = Heterogeneous
$^{*}p < 0.10$ $^{**}p < 0.05$ $^{***}p < 0.01$

competencies that closely matched their own – although, on average, there was a strong pattern of homogeneity between firm and partner capabilities.

The above results are consistent with the VoC thesis and strongly support the argument developed in this study. Firms from both countries appear to be guided by the specific sets of institutional arrangements that influence their actions and interactions in unique ways. Because these distinct institutional settings are accompanied by their own internal logics, it was expected that firms from both countries would generally seek out firms that reinforced rather than diluted their existing core competencies. For the reasons discussed above, LMEs tend to be slightly less "pure" in this sense because their institutional make up allows for greater within-system diversification (Boschma & Capone, 2015). Overall, however, the patterns identified above seem to suggest that firms look for partners at trade fairs that will help strengthen pre-existing economic and technological specializations. To the extent that such patterns of inter-firm collaboration are routinely followed, an imminent shift from divergence to convergence in the technological specializations of firms and the countries in which they operate seems unlikely.

It is important to note, however, that seemingly minor amendments, revisions, or additions to the existing rules governing certain industries or technological domains can, over time, bring about a more substantial change. In reference to what they call "layering", Mahoney and Thelen (2010, 17) suggest that substantial institutional change can occur "if amendments alter the logic of the institution or compromise the stable reproduction of the original 'core'". Because processes of layering do not actually remove the original rules, but simply add new rules alongside them, this type of change process may go largely unnoticed or run up against less resistance from those interested in maintaining the status quo (Mahoney & Thelen, 2010). As such, identifying and tracing such small changes while they are still in the process of accumulating can be extremely difficult. The garbage-can analysis presented in the previous chapter provides some clues as to how such small shifts might be captured. Yet, determining their ultimate impact still requires a much longer analytical timeline than what is presented here.

6.6 Conclusion

This chapter provided empirical evidence that differences in the national institutional contexts in which firms are embedded fundamentally shape their interactions with other firms and their search for technical solutions at trade fairs. This was demonstrated through a quantitative analysis of: (i) the information sources firms rely on to develop technological and industry know-how; (ii) the knowledge flows that take place between firms at trade fairs; (iii) the specific competencies that are seen to be central to the operations of the firms themselves; and, (iv) the competencies firms look for in potential partners during the trade fair.

With respect to information sources, the most striking finding was the difference between German and U.S. firms regarding their use of Internet-based information sources, with online platforms being viewed as significantly less important by German firms compared to their U.S. counterparts. Although not directly related to the study's core research questions, this finding resonates with the VoC approach, as the institutional conditions of LMEs tend to promote switchable assets and radical innovation (Hall & Soskice, 2001). It was argued that such a focus encourages firms to draw upon a wider variety of information sources, with publicly-available sources associated with the high-speed flows of online information being especially popular. In contrast, the particular institutional arrangements of CMEs tend to encourage German firms to engage in incremental innovations and make continuous improvements to high quality and customized products (Hall & Soskice, 2001). As such, it was expected that German firms would rely less on online information sources and put more energy into a few trusted sources of information based on long-term collaborative relationships.

As for knowledge flows, the above analysis discovered significant cross-national differences in where and what types of knowledge firms receive and what they transmit at trade fairs. With regards to both the sources of knowledge for firms and the recipients of firms' knowledge during trade fairs, there was a clear difference between the two countries regarding the emphasis placed on suppliers. Knowledge flows to and from suppliers at trade fairs were viewed as significantly more important for U.S. than German firms. This difference can be linked to the specific institutional frameworks of each capitalist variety. Firms in CMEs, like Germany, are encouraged to maintain long-term relationships with their suppliers, whereas firms in LMEs like the U.S. tend to choose their suppliers in response to more immediate and changing needs (Hall & Soskice, 2001). It was, thus, argued that focusing on potential suppliers during the trade fair might be less relevant for German firms, as long-term supplier relationships with existing suppliers would be more pertinent to their operations. For German firms, trade fairs are likely a key vehicle for knowledge exchange with other actors, such as potential new partners in research. But it may be regarded as a less valuable way of sharing knowledge with existing suppliers who are already deeply connected to such knowledge flows through other organized channels.

Finally, in terms of the core competencies of firms and partners, the analysis showed striking and significant differences between firms from Germany and the U.S. with respect to low-cost focus, constantly changing product line, and customized production. With respect to low-cost focus, U.S. firms were significantly more likely than their German counterparts to have core competencies in this area. Similarly, U.S. firms reported significantly higher competencies in the area of constantly changing product line. From the VoC perspective, these differences are well founded and can be linked to the embeddedness of U.S. firms in a market-based system that lends itself to rapid reconfigurations and the adoption of new technologies through short-term labour contracts, a mobile workforce, and limited worker involvement in, for example, wage negotiations (Hall & Soskice, 2001). Such conditions provide strong incentives for U.S. firms to put their energy into rapid innovations and constantly changing product lines, while greater labour market flexibility in such settings, along with a less specialized and highly mobile workforce encourages the maintenance of low-cost production systems (Hall & Soskice, 2001; Ruhs, 2013).

Aside from low-cost focus and constantly changing product line, customized production represents another important area of difference between the two countries, with German firms indicating much higher core competencies than their U.S. counterparts. This result also speaks to the core assumptions of the VoC model: CMEs, such as Germany, are characterized by strong vocational training systems, a highly specialized workforce, as well as enduring, collaborative, trust-based arrangements between firms, their employees, owners, and suppliers. These conditions tend to support incremental innovations and ongoing quality improvements in the area of customized production (Deeg, 1999; Hall & Soskice, 2001).

Further statistical analyses, including multivariate linear regression, revealed that the cross-national differences identified above remain significant even when the independent effects of other factors, such as firm size, age, and sector, are taken into account. Moreover, the analysis demonstrated that the core competencies of firms were closely matched by the competencies they searched for in potential partners. This pattern of convergence between firm and partner competences was observed in both countries, but was slightly more apparent in the case of German firms, which fits with differences in the degree of specialization/diversification associated with the different capitalist models. The overarching finding, however, was that firms from both Germany and the U.S. look for partners at the trade fair that closely match or reinforce their pre-existing central capabilities.

The results summarized above support the qualitative findings presented in Chap. 5 regarding the garbage-can-like processes of firms' search and interaction at trade fairs and create a broader empirical basis for these findings by linking them to more general, macro-level patterns of difference across distinct capitalist varieties. As will be discussed in the final chapter, the combined qualitative and quantitative findings have important implications for the study of capitalist diversity as well as current understandings of the role of international trade fairs in mediating global knowledge flows across geographic and political-economic boundaries.

References

Aarstad, J., Ness, H., & Haugland, S. A. (2015). Innovation, uncertainty, and inter-firm shortcut ties in a tourism destination context. *Tourism Management, 48*, 354–361.

Akkermans, D., Castaldi, C., & Los, B. (2009). Do "liberal market economies" really innovate more radically than coordinated market economies? Hall and Soskice reconsidered. *Research Policy, 38*, 181–191.

Aldrich, H. E., & Auster, E. R. (1986). Even dwarfs started small: Liabilities of age and size and their strategic implications. *Research in Organizational Behavior, 8*, 165–198.

Bathelt, H. (2017). Trade fairs and innovation. In H. Bathelt, P. Cohendet, S. Henn, & L. Simon (Eds.), *The Elgar companion to innovation and knowledge creation* (pp. 509–522). Edward Elgar.

Bathelt, H., Golfetto, F., & Rinallo, D. (2014). *Trade shows in the globalizing knowledge economy*. Oxford University Press.

Boschma, R., & Capone, G. (2015). Institutions and diversification: Related versus unrelated diversification in a varieties of capitalism framework. *Research Policy, 44*, 1902–1914.

Carayannopoulos, S. (2009). How technology-based new firms leverage newness and smallness to commercialize disruptive technologies. *Entrepreneurship Theory and Practice, 33*(2), 419–438.

Coleman, A. (2014). What Germany can teach the rest of Europe about twitter. *Forbes*, July 10. Retrieved June 2018, from https://www.forbes.com/sites/alisoncoleman/2014/07/10/what-germany-can-teach-the-rest-of-europe-about-twitter/#c0e99995d6f8

Crouch, C., & Streeck, W. (1997). *Political economy of modern capitalism: Mapping convergence and diversity*. Sage.

Dawes, J. (2008). Do data characteristics change according to the numbers of scale points used? *International Journal of Market Research, 50*(1), 61–77.

Deeg, R. (1999). *Finance capitalism unveiled: Banks and the German political economy*. University of Michigan Press.

Deeg, R., & Jackson, G. (2007). Towards a more dynamic theory of capitalist variety. *Socio-Economic Review, 5*, 149–179.

Dillman, D. A., Smythe, J. D., & Christian, L. M. (2009). *Internet, mail and mixed-mode surveys: The tailored design method*. Wiley.

Fagerberg, J. (1987). A technology gap approach to why growth rates differ. *Research Policy, 16*(2–4), 87–99.

Fagerberg, J. (2005). Innovation: A guide to the literature. In J. Fagerberg, D. C. Mowery, & R. R. Nelson (Eds.), *The Oxford handbook of innovation* (pp. 1–26). Oxford University Press.

Fink, A. (1995). *How to ask survey questions*. Sage.

Foddy, W. (1994). *Constructing questions for interviews and questionnaires: Theory and practice in social research* (Vol. 80, p. 382). Cambridge University Press.

Freeman, G. P., & Mirilovic, N. (Eds.). (2016). *Handbook on migration and social policy*. Edward Elgar Publishing.

Gavetti, G., & Levinthal, D. A. (2000). Looking forward and looking backward: Cognitive and experimental search. *Administrative Science Quarterly, 45*(1), 113–137.

Gilsing, V., Lemmens, C. E., & Duysters, G. (2007). Strategic alliance networks and innovation: A deterministic and voluntaristic view combined. *Technology Analysis and Strategic Management, 19*(2), 227–249.

Grant, R. M., & Baden-Fuller, C. (2004). A knowledge accessing theory of strategic alliances. *Journal of Management Studies, 41*, 61–84.

Guider, C. A. (2007). *The effects of size and structure of government on economic growth: A comparison of European models with the Anglo-American model*. Dissertation. Retrieved June 2017, from https://aquila.usm.edu/cgi/viewcontent.cgi?referer=&httpsredir=1&article=2344&context=dissertations.

Hall, P. A., & Gingerich, D. W. (2004). *Varieties of capitalism and institutional complementarities in the macroeconomy: An empirical analysis* (MPIfG discussion paper 04/5). Cologne: Max Planck Institute for the Study of Societies.

Hall, P. A., & Soskice, D. (2001). An introduction to varieties of capitalism. In P. A. Hall & D. Soskice (Eds.), *Varieties of capitalism: The institutional foundations of comparative advantage (pp. 1–68)*. Oxford University Press.

Hall, P. A., & Thelen, K. (2009). Institutional change in varieties of capitalism. *Socio-Economic Review, 7*(1), 7–34.

Hancké, B. (Ed.). (2009). *Debating varieties of capitalism: A reader*. Oxford University Press.

Hancké, B., & Herrmann, A. M. (2007). Wage bargaining and comparative advantage in EMU. In B. Hancké, M. Rhodes, & M. Thatcher (Eds.), *Beyond varieties of capitalism: Conflict, contradictions and complementarities in the European economy* (pp. 122–146). Oxford University Press.

Hite, J. M., & Hesterly, W. S. (2001). The evolution of firm networks: From emergence to early growth of the firm. *Strategic Management Journal, 22*(3), 275–286.

Jackson, G., & Deeg, R. (2008). Comparing capitalisms: Understanding institutional diversity and its implications for international business. *Journal of International Business Studies, 39*(4), 540–561.

Kitschelt, H. (1991). The 1990 German federal election and the national unification: A watershed in German electoral history? *West European Politics, 14*(4), 121–148.

Kitschelt, H., Lange, P., Marks, G., & Stephens, J. (Eds.). (1999). *Continuity and change in contemporary capitalism*. Cambridge University Press.

Kocka, J. (1984). Craft traditions and the labor movement in nineteenth century Germany. In P. Thane, G. Crossick, & R. Floud (Eds.), *The power of the past: Essays for Eric Hobsbawm*. Cambridge University Press.

Lee, J. A., & Soutar, G. (2010). Is Schwartz's value survey an interval scale, and does it really matter? *Journal of Cross-Cultural Psychology, 41*(1), 76–86.

Lichtenberg, F. R. (1992). *R&D investment and international productivity differences* (NBER Working paper, No. 4161).

Linnarson, H. (2005). Patterns of alignment in alliance structure and innovation. *Technology Analysis & Strategic Management, 17*(2), 161–181.

Mahoney, J., & Thelen, K. (2010). A theory of gradual institutional change. In J. Mahoney & K. Thelen (Eds.), *Explaining institutional change: Ambiguity, agency, and power* (pp. 1–37). Cambridge University Press.

Pearse, N. (2011). Deciding on the scale granularity of response categories of Likert type scales: The case of a 21-point scale. *The Electronic Journal of Business Research Methods, 9*(2), 159–171.

Pisano, G. P. (1994). Knowledge, integration, and the locus of learning: An empirical analysis of process development. *Strategic Management Journal, 15*(S1), 85–100.

Quintana-García, C., & Benavides-Velasco, C. A. (2011). Knowledge organisation in R&D alliances: Its impact on product innovation. *Technol Anal Strateg Manage, 23*(10), 1047–1061.

Rinallo, D., & Golfetto, F. (2011). Exploring the knowledge strategies of temporary cluster organizers: A longitudinal study of the EU fabric industry trade shows (1986–2006). *Economic Geography, 87*(4), 453–476.

Ruhs, M. (2013). *The price of rights: Regulating international labor migration*. Princeton University Press.

Ruhs, M. (2016). Theorizing labor immigration policies: Openness, skills, and rights. In G. P. Freeman & N. Mirilovic (Eds.), *Handbook on migration and social policy* (pp. 435–458). Edward Elgar Publishing.

Streeck, W. (1989). Skills and the limits of neo-liberalism: The enterprise of the future as a place of learning. *Work, Employment and Society, 3*(1), 89–104.

Streeck, W. (1996). Lean production in the German automobile industry: A test case for convergence theory. In S. Berger & R. Dore (Eds.), *National diversity and global capitalism* (pp. 138–170). Cornell University Press.

Streeck, W., & Thelen, K. (2005). Introduction: Institutional change in advanced capitalist economies. In W. Streeck & K. Thelen (Eds.), *Beyond continuity: Institutional change in advanced capitalist economies* (pp. 1–39). Oxford University Press.

Thelen, K. (2004). *How institutions evolve: The political economy of skills in Germany, Britain, the United States, and Japan*. Cambridge University Press.

Ulku, H. (2004). *R&D, innovation and economic growth: An empirical analysis* (IMF working paper 185). International Monetary Fund.

Yamamura, K., & Streeck, W. (Eds.). (2003). *The end of diversity? Prospects for German and Japanese capitalism*. Cornell University Press.

Yang, H., Steensma, H. K., & Phelps, C. (2010). Learning from what others have learned from you: The effects of knowledge spillovers on originating firms. *Academy of Management Journal, 53*(2), 371–389.

Chapter 7
Conclusion: Ongoing Specialization or Convergence of Capitalist Varieties?

7.1 Introduction

In this chapter, we revisit the three central research questions that have motivated this study. First, what is the nature of the search and interaction processes of firms at international trade fairs, specifically regarding knowledge flows related to technological innovation? Second, what is the potential significance of these events in the context of the global political economy and in relation to recent theories regarding the nature and dynamics of institutional change? And, third, to what extent do inter-firm interactions at trade fairs influence future patterns of technological production? Do such patterns lead towards processes of convergence or divergence in national technological specializations? In the next three sections, I summarize the insights derived from the study with respect to each of these questions. In recognition of the study's limitations, the fourth section identifies some potentially fruitful avenues for further research on this topic. The fifth section discusses the theoretical and practical implications of the book, paying particular attention to implications for the field of political science and the practice of public policy and policy-making at different levels. Finally, the chapter concludes by exploring the impacts of the COVID-19 pandemic on the trade fair industry and reflecting on the future role of international trade fairs in helping firms navigate the complex and continuously changing global political-economic context in which they operate.

7.2 Trade Fairs as "Organized Anarchies" and Garbage-Can Processes of Search and Interaction

Drawing on the garbage-can model (Cohen et al., 1972), as well as key ideas from the search and behavioural economics literatures, the research presented in this book sheds light on the nature of firms' search and information processes at trade

R. Gibson, *International Trade Fairs and Inter-Firm Knowledge Flows*,
https://doi.org/10.1007/978-3-031-20557-6_7

fairs, specifically with respect to how new technological innovations – or "solutions" – are identified. I argue that international trade fairs, like some other complex organizational settings, reflect the three basic features of an "organized anarchy" – namely, (i) problematic preferences, (ii) unclear technologies or processes, and (iii) fluid participation (Cohen et al., 1972). Within this complex organizational setting, the findings from the qualitative analysis, in Chap. 5, show that the decision-making processes of firms rarely follow a rational or linear path. Instead, they more closely resemble "garbage-can" decision-making processes – or processes by which largely independent streams of problems, solutions, and agents are brought together and move from "one choice opportunity to another" in such a manner that the very nature of the decision itself, the amount of time it takes to arrive at that decision, and the particular problem it addresses all hinge upon "a relatively complicated intermeshing of elements" (Cohen et al., 1972, 16).

The study found that while the technological search processes of firms at trade fairs were clearly inclined towards gathering new knowledge – be it related to products, technologies, the industry, the market, or management practices – these search processes were rarely motivated by a clearly-defined or pre-established set of goals. Instead, in the organized anarchy of the trade fair, processes of search and interaction are carried out by "boundedly rational" economic actors who frequently resort to "satisficing" (Simon, 1955, 1956) in order to make decisions at these events. Like the global economy itself, trade fairs were often viewed by firms as complex, fluid, and uncertain environments. Consequently, many of the firms in the study indicated that they would come to the trade fair with multiple loosely defined goals or a more emergent approach regarding what they hoped to find at the trade fair; predefined goals or clear objectives were much less common. These results support previous findings in the technological search literature, particularly the argument that firms tend to engage more in exploitative than explorative search processes (Pavitt, 1998; Helfat, 1994; Stuart & Podolny, 1996; Tripsas & Gavetti, 2000; Laursen, 2012). That is, firms tend to be more involved in searching for solutions that are closely related to their current production contexts and pre-existing bases of knowledge (i.e., exploitative search) than engaging in a deliberate effort to move away from these pre-established foundations (i.e., explorative search).

In general, this study found that experiential, learning-by-doing processes of technological search were more common than cognitive, learning-before-doing processes. Yet, the search and interactions of firms assumed different forms, such as relying on heuristic short-cuts. For example, learning vicariously from other firms – whereby firms observe the actions and attendant outcomes of other organizations and then adopt or mirror the most successful of these behaviours (Yang et al., 2010) – was a type of search heuristic that featured prominently in the empirical findings. As a representative of a digital manufacturing firm explained, "I will walk around and see what people are doing, see what they have decided. [...] The signage here, we found at a show a few years ago. Somebody was doing something very similar to it and for me it was a possibility, so we put it together."

Despite evidence of different approaches to technological search and the heterogeneity of choices being made at trade fairs, the qualitative analysis strongly suggests that the processes of search and inter-firm interaction at these events are invariably shaped by the firm's pre-existing organizational routines, production contexts, and knowledge bases – all of which are deeply rooted in the institutional make-up of the firm's home country. The next two sections further explore the consequences of this finding, first, in relation to recent theories of incremental institutional change and, second, in relation to ongoing debates regarding the divergence or convergence of capitalist varieties.

7.3 International Trade Fairs as Key Nodes in the Global Political Economy

Building on recent conceptualizations of trade fairs as "temporary clusters" (Maskell et al., 2006; Bathelt et al., 2014), this study contributes new empirical evidence regarding the potential significance of these events in processes of knowledge circulation by adopting a specific focus on the political-economic dynamics of such knowledge flows in relation to the technological development paths of distinct capitalist models. In keeping with the knowledge-based view of trade fairs (Bathelt et al., 2014), the study suggests that international trade fairs represent key nodes in the global political economy that drive processes of inter-firm learning, knowledge exchange, and innovation. Trade fairs bring together both leading and less well-known actors from a given industry or technology field mainly for the purpose of exchanging knowledge about new trends and innovations in their respective industry. For a designated period of time, trade fair exhibitors are deeply immersed in a setting that features rich and varied knowledge flows. These knowledge flows have the ability to cut across distinct geographic and political-economic boundaries, creating multiple opportunities for interaction, observation, and exchange among the trade fair participants. Such flows are unique to the trade fair setting and can serve to link markets and technologies with specific institutional and political conditions.

The combined qualitative and quantitative findings of this study suggest that trade fairs may play an important role in shaping global economic dynamics by mediating the negotiation of differences between capitalist economies regarding their technological specializations and, possibly, the institutional practices and knowledge bases that underpin them. Whereas the qualitative findings summarized above suggest that the institutional structures of the firm's home base may play a pivotal role in guiding the nature of its search activity at trade fairs, the quantitative analysis extends the basis for these findings by identifying broad differences in the behaviours of German and U.S. firms at these events. The quantitative analysis identified significant cross-national differences in the following areas: (i) information sources firms relied on to develop technological and industry know-how; (ii) knowledge flows that took place between firms at trade fairs; (iii) core competencies that were considered central to the operations of the firm; and (iv) core competencies firms looked for in potential partners during the trade fair. Additional statistical tests, including multivariate linear regression, revealed that these cross-national differences remained significant even when the independent effects of other factors, such as firm size, age, and sector, were taken into account. Further, the cross-national differences identified in this study are largely consistent with the expectations of the VoC literature regarding the character of LMEs versus CMEs.

In addition to deepening our understanding of the potential role of trade fairs in the global political economy, this study also aims to demonstrate the potential utility of a focus on trade fairs within the context of recent debates regarding the development of more dynamic accounts of capitalist diversity (Deeg & Jackson, 2007; Jackson & Deeg, 2008) as well as theoretical developments on the topic of incremental institutional change (Thelen, 2004; Streeck & Thelen, 2005; Mahoney & Thelen, 2010). First, the study argues that trade fairs constitute a potentially significant – yet still largely under-examined – source of economic dynamism in the current globalizing knowledge economy. The reason for this, as developed in Chap. 3, is that international trade fairs embody the four building blocks of what Arvanitidis and Petrakos (2011) call "knowledge-driven economic dynamism". These four building blocks are: (i) human capital; (ii) innovation ability; (iii) information access; and (iv) economic performance.

Second, while the study's empirical results seem to point more towards continued stability than change in the technological specialization patterns of firms, applying the garbage-can model to the context of trade fairs introduces the possibility of a more dynamic interpretation of the way in which the actions and interactions of firms during these events might be associated with future technological development paths. The garbage-can model's notion of "unclear technologies", for instance, describes organizational choice as involving processes of trial-and-error, learning from past mistakes, and "pragmatic inventions of necessity" (Cohen et al., 1972, 1). In the trade fair context, such "pragmatic inventions of necessity" may relate to the ongoing yet gradual adaptations of a firm's practices in response to dynamic markets. Some examples might include: the adoption of a new technology or application, retiring a product that has lost its relevance in the industry, or establishing linkages to new external markets when existing ones become less viable. While it is expected that the nationality of the firm will continue to play a decisive role in shaping how that firm adapts, over time such incremental adjustments could influence the development pathways of firms and possibly even the larger technology fields or industries in which they are embedded. As such, firms participating in international trade fairs might inadvertently become "change agents" (Mahoney & Thelen, 2010) by adapting their practices in subtle ways that, over time, may collectively amount to a more transformative shift in institutional practices.

Finally, the multi-level framework described in the preceding chapters was developed in response to Deeg and Jackson's (2007) argument that introducing more dynamism into the VoC literature requires proceeding simultaneously along three distinct levels of analysis – micro, meso, and macro. In the present study, the micro-level consists of firms' search and interaction processes at trade fairs; the macro-level analysis centres on capitalist varieties and the convergence-divergence debate; and international trade fairs are positioned as critical meso-level structures connecting the micro- and macro-levels by enabling and mediating the exchange of technological knowledge between participating firms and across capitalist varieties.

7.4 Convergence or Divergence in Patterns of Technological Specialization?

Based on discussions in the literature on comparative capitalisms, this study has looked at how the institutional foundations of capitalist varieties might shape inter-firm interactions at trade fairs and, in turn, guide future patterns of technological production – first, at the level of the firm, but ultimately more broadly at the level of the national state. This investigation involved reflecting on the long-standing debate regarding the convergence or divergence of distinct varieties of capitalism, specifically with respect to patterns of technological specialization. Given the limitations in the data – most notably, the absence of longitudinal measures – it was not possible to establish a clear causal relationship between inter-firm interactions at trade fairs and the future development pathways of firms and their underlying national political economies. However, the study's combined qualitative and quantitative findings enabled me to speak to both the micro processes of firm behaviour at trade fairs as well as more general tendencies regarding differences between German and U.S. firms. Using the quantitative data (Chap. 6) as a means to broadening the basis of the qualitative findings (Chap. 5), it was, therefore, possible to develop a compelling narrative about the likelihood of convergence or divergence in national patterns of technological specialization.

Without repeating the points made in the previous section regarding how the study's findings might add value to theories of institutional change, the qualitative analysis clearly illustrates that the search processes of firms at trade fairs are strongly guided by the specific production contexts and knowledge bases that are associated with the country in which they operate. Further, the quantitative analysis provides strong evidence that U.S. firms differ in significant ways from their German counterparts with respect to the core competencies they see as central to the operations of their firm as well as those they look for in potential partners at the trade fair. U.S. firms, for example, were significantly more likely than German firms to have core competencies in low-cost focus and constantly changing product line. German firms, on the other hand, indicated significantly higher core competencies in customized production than their U.S. counterparts. Finally, for both U.S. and German firms, there was generally a high level of homogeneity between the core capabilities of firms and the partners they looked for at trade fairs – for instance, firms that rated customized production as a very important competence for their firm tended to also rate customized production highly in their partners. All of this suggests that the nature of inter-firm interactions at trade fairs tends to reinforce rather than undermine existing patterns of technological specialization, thereby setting firms and the capitalist varieties to which they belong on a pathway towards continued divergence, not convergence.

7.5 Implications of a Focus on Trade Fairs for Political Science

This study represents the first systematic empirical analysis linking inter-firm interactions at international trade fairs to distinct varieties of capitalism. As such, this section considers the implications of a focus on international trade fairs for the field of political science. This includes implications for the comparative study of capitalist varieties as well as implications for public policy and policy-making in the context of the global knowledge-based economy. Such implications are discussed from both a theoretical and practical vantage point.

7.5.1 *Theoretical Implications*

As previously noted, international trade fairs have been largely neglected in the field of political science. In the political economy literature, for instance, trade fairs usually only receive passing reference and, for the most part, such references tend to be historical in nature (Frieden, 2006; O'Brien & Williams, 2016). In their contemporary form, trade fairs are all but absent in this literature. Indeed, many of the key political economy textbooks – including those with a global focus – overlook trade fairs entirely (see, for example, Gilpin, 2001; Ravenhill, 2008; Oatley, 2016). This study provides compelling evidence as to why trade fairs should be taken more seriously, particularly within the realm of comparative political economy studies. More specifically, within the context of ongoing discussions regarding the mechanisms of stability versus change in capitalist diversity (e.g., Hall & Soskice, 2001; Deeg & Jackson, 2007; Haddow, 2008; Hall & Thelen, 2009), this book suggests that international trade fairs can play a pivotal role in connecting different models of capitalism, and that the nature of inter-firm interactions at these events closely reflects the specific institutional arrangements of the national economic systems in which firms' operate. From this, it has been argued that international trade fairs might be linked to the persistence and continued specialization of distinct varieties of capitalism; yet further research of a longitudinal nature is needed to clarify such linkages.

With respect to recent efforts to develop more dynamic theories of capitalist diversity (Deeg & Jackson, 2007), including important theoretical advances in the study of incremental institutional change processes (Streeck & Thelen, 2005; Mahoney & Thelen, 2010), this study reveals that institutional path dependence and processes of knowledge contextualization and localization remain strong, even within highly complex organizational settings made up of global actors that are engaged in seemingly random, non-linear decision-making processes. Yet, identifying and tracing the more gradual forms of institutional change that might be at play within the trade fair setting require a longer-term perspective than what has been possible here. Moreover, as Mahoney and Thelen (2010) point out, the characteristics of the political context and institutional form help determine the type of "change agent" that is likely to participate and thrive in a given institutional context, as well as the kinds of "change strategies" such participants will be inclined to pursue. It seems quite conceivable that future changes in the political-economic landscape – such as a global pandemic or a protectionist turn in global economic policy – might influence which actors participate in international trade fairs as well as the ideas and preferences they bring with them. An argument along these lines would fit nicely with the garbage-can model's view of "fluid participation" and "unclear preferences" in organized anarchies.

While the empirical focus of this study is on the U.S. and Germany, international trade fairs have the capacity to play a decisive role in shaping economic development trajectories and transition dynamics of less-developed political economies (Bathelt et al., 2014). This has already been empirically demonstrated in recent work focused on trade fair ecologies in the Asia-Pacific region (Bathelt & Zeng, 2014). The findings from the present study also suggest that international trade fairs may be an important vehicle for less-developed economies, as leading trade fairs provide multiple opportunities for firms to make cross-system connections and take part in the rich array of knowledge flows through a variety of different search processes, such as learning-by-doing and vicarious learning. In this way, international trade fairs provide crucial insights for firms regarding how they might best position themselves within the context of global economic competition.

7.5.2 Policy Implications

In addition to the study's theoretical implications, an important motivation of this work is to offer insights that may help inform governments, policy-makers, and other actors involved in ensuring the economic growth and competitiveness of local firms and the national and regional states in which they operate. From this perspective, this section discusses implications for policy and policy making related not only to export promotion, but also to learning opportunities for firms in innovation and regional/national economic development.

In developing the argument that international trade fairs represent key nodes in the global political economy, this study suggests that trade fairs are an important vehicle through which firms can remain competitive in the face of ongoing changes in the global political economy. International trade fairs have long been viewed as a valuable tool for export promotion in a broad sense, resulting in an increase in sales to other countries or regions. Recent research (Tinits & Fey, 2022), for example, explores the efficacy of trade fairs not only in relation to other export support mechanisms – strategic export planning, participation in ministerial visits – but also in terms of their temporal sequencing of export support mechanisms. This research finds that "exports are best facilitated by first receiving support to attend a trade fair, then developing an export plan" (Tinits & Fey, 2022, 1). Despite the clear value of trade fairs from an export promotion standpoint, the present study and other recent work (Bathelt et al., 2014) suggest that these events offer unique access to rich and varied knowledge flows about external markets and other relevant opportunities that could be equally, if not more, valuable in terms of successfully adapting to ongoing changes in global markets, especially in the context of the global knowledge economy.

In addition to providing important opportunities to gain access to new markets and stimulate trade, this study also demonstrates how trade fairs are used to showcase the core competencies of exhibiting firms. In doing so, trade fairs represent critical platforms for networking and relationship-building, allowing firms to search for complementary capabilities in potential business partners. In this way, trade fairs should be viewed as an important mechanism for strengthening existing comparative advantages as well as supporting government initiatives to foster new innovations in relevant areas of technological specialization. Trade fairs provide valuable "benchmarks" for exhibiting firms in developing innovative and competitive products and technologies (Bathelt et al., 2014). They encourage firms to innovate in different ways (e.g., radical/incremental) and in different areas of competence that build on and reinforce existing strengths. Choosing the "right" or most valuable trade fairs to attend as well as having sufficient opportunities to participate in relevant non-local events are, therefore, vitally important in constructing and maintaining the innovation capabilities of firms.

The research presented in this book supports findings from other studies (Bathelt et al., 2014, 275) which demonstrate that "trade shows are important learning platforms that can assist firms in developing the technological and market-related knowledge necessary to innovate based on observing and interacting with non-local partners, competitors, and customers." Yet, in providing more insight into the precise nature and types of search processes that tend to occur at these events, innovation policies might be designed and employed more effectively to support or strengthen, for example, the knowledge contextualization activities of firms at trade fairs. Beyond the explicit innovation contests or awards that are often part of international trade fairs, this study suggests that the potential for trade fairs to serve as a vital policy tool to support innovation may often be found in the less overt or planned routes where firms happen upon a "solution" to an existing problem in their local production context or something which could be adapted to suit their own technological setting.

What all of this suggests, from a practical or policy perspective, is that firms need to be well supported – in financial and other ways – to actively participate in relevant, often high-profile international trade fairs, which may be held in distant locations. While government support for overseas trade fair participation is longstanding and widespread, the nature and organization of such programs vary considerably across national economic systems (Hauswirth, 2006). In Germany, for instance, trade fairs have long been an important part of the national political economy. In this context, trade fairs operate in a stable, well-established and coordinated policy environment that is supported by institutions specifically designed for the promotion of the trade fair industry (e.g., AUMA). The trade fair industry in North America is a much newer development, but its growth has been particularly evident in certain parts of the United States, such as Las Vegas, Nevada. It is important for governments and policy-makers in all countries to consider the policies and programs currently in place to support such events, as well as the implications of these policy frameworks with respect to their implications for both national and global economic competitiveness.

Policy recommendations in this area, therefore, might include: first, financial support to allow firms – particularly small and medium-sized firms, which may be less well established and lack the necessary resources – to participate in international trade fairs that feature, for instance, industry or technology leaders, potential partners with complementary competencies, or new markets for innovation; second, training or instructional programs for firms that do participate in trade fairs to ensure that such events are used in a way that maximizes learning and innovation, rather than stressing marketing or promotional activities above all else. Part of such a training might involve highlighting the unique characteristics of international trade fairs discussed in this research, such as the conceptualization of trade fairs as organized anarchies or the relationship between trade fairs and knowledge-driven economic dynamism – including its core components: innovation capacity, information access, human capital, and economic performance; a third recommendation, which relates to both resources and the shift in priorities noted above, might involve the number and nature of firm representatives sent to participate in international trade fairs. In order to have sufficient time to engage in search activities, firms need to have enough staff engaged in trade fair participation – that means, at minimum, allowing for a sufficient and consistent presence at the firm's own trade fair booth while, at the same time, having some staff free to explore other exhibits. The specific positions of the staff members participating in the trade fair may also matter depending on what type of learning or innovation is needed – a firm's sales representative, for example, would be well-suited to identifying innovations in sales or marketing, yet may be less able to recognize "solutions" or potential innovations in other areas compared to, say, a designer, technical lead, engineer, or so on.

7.6 Thoughts on a Future Research Agenda

As expressed in the implications section above, further research is needed to identify the precise mechanisms by which international trade fairs connect businesses on a global scale and influence technological developments at different levels – e.g., international, national, sub-national, and sectoral. This study has taken an initial step towards elucidating the relationship between trade fairs and the dynamics of capitalist varieties. Yet, there remain many open questions. Building on the findings presented in this book, a promising area of future research might involve the adoption of a longitudinal approach that can connect, in a more systematic way, the nature of inter-firm interactions at trade fairs to empirically observable patterns of change and stability in the technological specialization patterns and national production contexts of firms. Such an analysis would be able to contribute more definitively to the broader VoC literature, including recent discussions regarding the more dynamic elements of capitalist diversity, which may involve more gradual changes in institutional practices (Deeg & Jackson, 2007; Hall & Thelen, 2009; Mahoney & Thelen, 2010).

As part of the abovementioned research agenda, or perhaps as a separate project, it would also be useful to engage in a more fine-grained treatment of the core competencies of firms and potential partners – for example, identifying the radical and incremental aspects of innovation focus and then linking these through a longitudinal lens to the development paths of different varieties of capitalism. Such an extension would build upon recent empirical work that challenges and refines the VoC's original interpretation of the radical-incremental innovation dichotomy (see, for example, Taylor, 2004, 2016; Akkermans et al., 2009).

A third area of research that could serve as a valuable foundation for future comparative work on this topic would be a detailed mapping of the policies and programs in place in different countries to support international trade fairs and trade fair participation. A study along these lines could form the basis of a comparative analysis of the role of international trade fairs as a policy tool for governments across different industries and sectors in areas such as export promotion, technological innovation, and economic development more generally.

7.7 Outlook: COVID-19 and the Future of International Trade Fairs

In the span of only 2 years, the COVID-19 pandemic has had a dramatic impact on the trade fair industry, as travel bans and strict health and safety measures were implemented worldwide in an effort to reduce the spread of the virus. In the case of Germany, for example, a country that accounts for close to 10% of the world's exhibition market, more than 70% of the 370 planned trade fairs in 2020 were cancelled as a result of the pandemic and its associated restrictions (AUMA, 2022a). In 2021, more than two-thirds of the German trade fair industry's 380 planned events were cancelled in the first half of the year (AUMA, 2022a). A similar scenario unfolded in countries around the world. According to UFI (the Global Association of the Exhibition Industry), the global trade fair industry contracted by a remarkable 68% in 2020 compared to the previous year, resulting in an estimated $370 billion (USD) loss in trade between exhibiting firms (UFI, 2021a).

In response to COVID-related restrictions as well as concerns about the potential of trade fairs to become superspreader events, trade fair organizers in many countries developed a variety of digital alternatives to traditional in-person events. In Germany, for instance, approximately 50–60 digital replacement events took place in 2020 and 2021 (AUMA, 2022a). According to a survey of 427 German exhibiting firms, conducted by AUMA (Association of the German Trade Fair Industry) in the fourth quarter of 2020, such digital events were recognized as an important substitute to conventional trade fairs; yet, for the majority of exhibiting firms, these online events did not represent a serious or permanent alternative to conventional in-person trade fairs (AUMA, 2021). To briefly illustrate: only 21% of survey respondents felt that virtual events represented a real alternative to physical in-person trade fairs; 17% of exhibiting firms indicated that they were using the digital replacement events, in 2020, because of trade fair cancellations, but that they would stop relying on these events once in-person trade fairs resumed. Nearly half (48%) of exhibiting firms indicated a preference for a mix of in-person and digital trade fair elements going forward, saying they would like to continue their reliance on physical trade fairs, but also make use of some of the digital supplements; and about 14% did not think that purely digital events could serve as an alternative, even in the absence of in-person trade fairs over the longer term (AUMA, 2021).

The above sentiments are generally supported by UFI's latest Global Barometer report (UFI, 2022),[1] which, in the context of the pandemic's impact on the global exhibition industry, explores possible trends in the future format of trade fairs. According to the report, 80% of respondents agreed with the statement that "COVID-19 confirms the value of face-to-face events", and there was a broad expectation that the industry would recover quickly (with 38% responding "Yes, for sure" and 42% "Most probably") (UFI, 2022, 15–16). Further, nearly three-quarters (73%) of respondents felt there has been a "A push toward hybrid events, more digital elements at events" (UFI, 2022, 15–16). Finally, only 10% agreed with the following statement: "Virtual events are replacing physical events" (UFI, 2022, 15–16).

Although purely digital events may not be viewed as a viable alternative for the majority of trade fair exhibitors, as noted above, there seems to be considerable momentum towards hybrid events, which involve a mix of in-person and online components. Indeed, many trade fair organizers have included hybrid solutions as a part of their future event strategies (Meetingspotlight, 2020; Carey, 2021; Jung & Lee, 2022; PCMA, 2022). For example, a recent survey probing the intentions of event organizers revealed that 73% of organizers plan to run a hybrid event in the future (Meetingspotlight, 2020). We are already seeing such hybrid events, as COVID-19 restrictions specifically related to trade fairs have eased or been dropped in many parts of the world and physical events have started to resume (AUMA 2022b). In the U.S., for example, the 52nd edition of the influential Consumer Electronics Show (CES) was changed from a purely virtual trade fair in 2021 to a hybrid event in 2022; it maintained a physical presence over a three-day period in Las Vegas, Nevada, while also offering a number of digital elements – including, livestreamed conference sessions, keynotes, and press conferences (CES, 2021a, b). Even before the disruptions caused by the pandemic, studies had begun to explore the interplay between conventional and virtual trade fairs, and the ways in which

[1] Since the financial crisis of 2008, the UFI Global Barometer has assessed the impact of global economic conditions on the trade fair and exhibitions industry. The latest report is based on a global survey of 401 participating firms from 53 countries and regions. It provides important updates on the impacts of COVID-19 and shares perspectives for the exhibition industry for the year 2022 (UFI, 2022).

physical and virtual elements of these events might complement one another (Bathelt & Turi, 2013; Gottlieb & Bianchi, 2017; Sarmento & Simões, 2019). Recent industry trends towards hybrid trade fairs will likely serve to stimulate more research on this topic. This will be important not only to inform the future development and design of such events, but also to better understand how such changes might influence, for example, the ways in which exhibitors connect with customers or search for new innovations or business partners at trade fairs.

During the completion of this book, the COVID-19 pandemic continues to evolve and play out differently in different parts of the world. While the full extent of its impacts on the global trade fair industry and the nature of the industry's recovery may not yet be fully realized, there are, at present, promising signs that the industry is starting to rebound. According to the Center for Exhibition Research (CEIR) First Quarter Index, for example, the U.S. business-to-business (B2B) exhibitions industry saw a significant drop in cancellation rates for physical in-person events in the first quarter of 2022 compared to the previous eight quarters (CEIR, 2022). To briefly illustrate, the cancellation rate for physical in-person events in the first quarter of 2022 was 9% compared to 91% in the first quarter of 2021 (CEIR, 2022). The Professional Convention Management Association's most recent COVID-19 recovery dashboard, based on a survey of 580 planners and suppliers, also points to signs of recovery: with more in-person and hybrid events being organized in the coming months, "the industry is solidly in return-to-in-person mode" (PCMA, 2022, 1). Similarly, the latest UFI Global Recovery Insights series (UFI, 2021b)[2] as well as its Global Barometer report (UFI, 2022) suggest optimism in the trade fair industry and a growing demand on behalf of exhibitors and visitors for live in-person events. When asked to identify the factors that would be most helpful in enabling exhibitions to "bounce back", 70% of firms pointed to the "Lifting of current travel restrictions", while about half of firms identified the following: "Readiness of exhibiting companies and visitors to exhibit again" (54%), "Mid-term visibility in terms of public policies, including travel restrictions" (49%), and "Lifting of current public policies that apply locally to exhibitions" (47%). Another 41% of firms suggested that "Financial incentive packages (leading to reduced costs for exhibitors)" would help support the "bounce back" of exhibitions (UFI, 2022, 5, 7).

Although the global trade fair industry appears to be gradually recovering from the disruptions caused by the COVID-19 pandemic, the nature of this recovery is expected to be both non-linear and variable across countries and regions. The nature and pace of recovery will also be influenced by new and emergent factors, including political conflict, deglobalization trends, economic and environmental crises, and interrupted supply chains. As part of a future research agenda, it will be important to closely examine the nature and dynamics of recovery of the trade fair industry across different geographies, political-economic systems, and sectors. It will also be

[2] The Global Recovery Insights series is a quantitative global survey of trade fair exhibitors and visitors, the latest of which was carried out during the summer of 2021. Collectively, the series has obtained 15,000 responses in 10 languages and represents trade fair participation in over 30 countries (UFI, 2021b).

interesting to observe how the virtual or digital aspects of trade fairs are further developed in the coming months and years, and how these aspects are incorporated into conventional trade fair settings. In the context of the present study, it will be particularly fascinating to see how such changes affect the dynamics of firms' search and interaction processes at trade fairs. While it is beyond the scope of the present study to engage in debates around the merits and shortcomings of virtual trade fairs, or their potential to replace in-person events (but see, for example, Lee-Kelley et al., 2004; Bathelt & Schuldt, 2010; Geigenmüller, 2010; Li, 2010; Gottlieb & Bianchi, 2017; Jung & Lee, 2022), based on the discussion above, there is reason to believe that physical trade fairs will retain a vital role in enabling inter-firm interaction and global knowledge flows in the foreseeable future. As Bathelt et al. (2014, 281) point out, trade fairs have proven to be remarkable resilient: "they have constantly renewed themselves and adjusted to changing economic, political, cultural, social, and technological conditions, despite repeated challenges and various periods of decline."

In light of recent and ongoing developments affecting the global political economy – including the Russia-Ukraine war, the COVID-19 pandemic, and the persistence of protectionist economic policies in multiple countries – it is important that governments and policy-makers in all countries consider the policies and programs currently in place to support international trade fairs, as well as the implications of these policy frameworks with respect to both national and global economic competitiveness. In the current political-economic climate, and surely in others to come, leading international trade fairs may increasingly be relied upon to help firms remain viable or even thrive in the face of such complex and dynamic conditions. Indeed, in the context of the pandemic, trade fairs have been seen as a "lifeline" for many firms in conducting their global business activities. Based on what we know of international trade fairs, including their resilience and adaptability over time, it seems likely that these events will continue to play an important role in helping the global economy recover from the impacts of COVID-19 as well as navigate other imminent and still unknown disruptions that lie ahead.

References

Akkermans, D., Castaldi, C., & Los, B. (2009). Do "liberal market economies" really innovate more radically than coordinated market economies? Hall and Soskice reconsidered. *Research Policy, 38*, 181–191.

Arvanitidis, P. A., & Petrakos, G. (2011). Defining knowledge-driven economic dynamism in the world economy: A methodological perspective. In P. Nijkamp & I. Siedschlag (Eds.), *Innovation, growth and competitiveness: Dynamic regions in the knowledge-based world economy* (pp. 15–39). Springer.

AUMA (Association of the German Trade Fair Industry). (2021). *Digital events: No permanent alternative for most exhibitors in Germany*. January 19, 2021. Retrieved May 8, 2022, from https://www.auma.de/en/media/reports/press-2021-02

AUMA (Association of the German Trade Fair Industry). (2022a). *Corona pandemic and trade fairs*. Retrieved May 1, 2022, from https://www.auma.de/en/media/reports/corona-virus-information

AUMA (Association of the German Trade Fair Industry). (2022b). *No COVID-19 entry requirements for trade fair participants*. October 6, 2022. https://www.auma.de/en/covid-19-pandemic/entry-requirements

Bathelt, H. & Schuldt, N. (2010). International trade fairs and global buzz, part I: Ecology of global buzz. *European Planning Studies, 18*(12), 1957–1974.

Bathelt, H. & Turi, P. (2013). Knowledge creation and the geographies of local, global, and virtual buzz. *Knowledge and the Economy, 5*, 61–78.

Bathelt, H., Golfetto, F., & Rinallo, D. (2014). *Trade shows in the globalizing knowledge economy*. Oxford University Press.

Bathelt, H., & Zeng, G. (2014). The development of trade fair ecologies in China: Case studies from Chengdu and Shanghai. *Environment and Planning A: Economy and Space, 46*(3), 511–530.

Carey, R. (2021). Hybrid meetings: What's new, what's next. *Meetingsnet*. Retrieved May 18, 2022, from https://www.meetingsnet.com/event-tech-virtual-meetings/hybrid-meetings-what-s-now-what-s-next

CEIR (Center for Exhibition Industry Research). (2022). *CEIR announces 2022 first quarter index results*. Retrieved June 2022, from: https://www.ceir.org/news/ceir-announces-2022-first-quarter-index-results/#:~:text=Excluding%20cancelled%20events%2C%20the%20Total,from%202019%20in%20Q4%202021

CES (2021a). CES to open with 2200+ in-person exhibitors. *CES Press Release*. Retrieved May 2022, from https://www.ces.tech/News/Press-Releases/CES-Press-Release.aspx?NodeID=9a9009e2-a501-4dc3-b328-0960e2a385d2

CES (2021b). Web Summit selected as digital platform provider. *CES Press Release*. Retrieved May, 2022, from https://www.ces.tech/News/Press-Releases/CES-Press-Release.aspx?NodeID=c062695f-48d1-4977-a7a5-407a8fb9486a

Cohen, M. D., March, J. G., & Olsen, J. P. (1972). A garbage can model of organizational choice. *Administrative Science Quarterly, 17*(1), 1–25.

Deeg, R., & Jackson, G. (2007). Towards a more dynamic theory of capitalist variety. *Socio-Economic Review, 5*, 149–179.

Frieden, J. A. (2006). *Global capitalism: Its fall and rise in the twentieth century*. Norton.

Geigenmüller, A. (2010). The role of virtual trade fairs in relationship value creation. *Journal of Business & Industrial Marketing, 25*(4), 284–292.

Gilpin, R. (2001). *Global political economy: Understanding the international economic order*. Princeton University Press.

Gottlieb, U. & Bianchi, C. (2017). Virtual trade shows: Exhibitors' perspectives on virtual marketing capability requirements. *Electronic Commerce Research & Applications, 21*, 17–26.

Haddow, R. (2008). How can comparative political economy explain variable change? Lessons for, and from, Canada. In L. A. White, R. Simeon, R. Vipond, & J. Wallner (Eds.), *The comparative turn in Canadian political science* (pp. 221–237). UBC Press.

Hall, P. A., & Soskice, D. (2001). An introduction to varieties of capitalism. In P. A. Hall & D. Soskice (Eds.), *Varieties of capitalism: The institutional foundations of comparative advantage* (pp. 1–68). Oxford University Press.

Hall, P. A., & Thelen, K. (2009). Institutional change in varieties of capitalism. *Socio-Economic Review, 7*(1), 7–34.

Hauswirth, I. A. (2006). *Effective and efficient organisations?: Government export promotion in Germany and the UK from an organisational economics perspective*. Springer.

Helfat, C. E. (1994). Evolutionary trajectories in petroleum firm R&D. *Management Science, 40*, 1720–1747.

Jackson, G., & Deeg, R. (2008). Comparing capitalisms: Understanding institutional diversity and its implications for international business. *Journal of International Business Studies, 39*(4), 540–561.

Jung, S., & Lee, J. (2022). Current and future influences of COVID-19 on the knowledge management function of conventions and exhibitions. *Service Business, 16*, 391–410. https://doi.org/10.1007/s11628-022-00488-7

Laursen, K. (2012). Keep searching and you'll find: What do we know about variety creation through firms' search activities for innovation? *Industrial and Corporate Change, 21*(5), 1181–1220.

Lee-Kelley, L., Gilbert, D. & Al-Shebabi, N. F. (2004). Virtual exhibitions: An exploratory study of Middle East exhibitors' dispositions. *International Marketing Review, 21*(6), 634–644.

Li, L. Y. (2010). Antecedents and effect of internet implementation for trade shows. *Journal of Business Industrial Marketing, 25*(4), 272–283.

Mahoney, J., & Thelen, K. (2010). A theory of gradual institutional change. In J. Mahoney & K. Thelen (Eds.), *Explaining institutional change: Ambiguity, agency, and power* (pp. 1–37). Cambridge University Press.

Maskell, P., Bathelt, H., & Malmberg, A. (2006). Building global knowledge pipelines: The role of temporary clusters. *European Planning Studies, 14*(8), 997–1013.

Meetingspotlight. (2020). *Stats: 95% of event organizers likely to run hybrid event in 2020*. Retrieved May 8, 2022, from https://www.meetingspotlight.com/article/stats-95-event-organizers-likely-run-hybrid-event-2020

Oatley, T. (2016). *International political economy*. Taylor & Francis.

O'Brien, R., & Williams, M. (2016). *Global political economy: Evolution and dynamics* (5th ed.). Palgrave Macmillan.

Pavitt, K. (1998). Technologies, products and organization in the innovating firm: What Adam smith tells us and Joseph Schumpeter doesn't. *Industrial and Corporate Change, 7*(3), 433–452.

PCMA (Professional Convention Management Association). (2022). *COVID-19 recovery dashboard – Survey dates: April 12–19, 2022*. Retrieved May 18, 2022, from https://www.pcma.org/wp-content/uploads/2022/05/0419-Dashboard-19-NT-2-CWW-2.pdf?utm_source=custom_redirect&utm_medium=web&utm_term=2022&utm_content=survey-23&utm_campaign=covid-19

Ravenhill, J. (2008). *Global political economy* (2nd ed.). Oxford University Press.

Sarmento, M. & Simões, C. (2019). Trade fairs as engagement platforms: The interplay between physical and virtual touch points. *European Journal of Marketing, 53*(9), 1782–1807.

Simon, H. A. (1955). A behavioral model of rational choice. *The Quarterly Journal of Economics, 69*(1), 99–118.

Simon, H. A. (1956). Rational choice and the structure of the environment. *Psychological Review, 63*(2), 129.

Streeck, W., & Thelen, K. (2005). Introduction: Institutional change in advanced capitalist economies. In W. Streeck & K. Thelen (Eds.), *Beyond continuity: Institutional change in advanced capitalist economies* (pp. 1–39). Oxford University Press.

Stuart, T., & Podolny, J. (1996). Local search and the evolution of technological capabilities. *Strategic Management Journal, 17*, 21–38.

Taylor, M. Z. (2004). Empirical evidence against varieties of capitalism's theory of technological innovation. *International Organization, 58*(3), 601–631.

Taylor, M. Z. (2016). *The politics of innovation: Why some countries are better than others at science and technology*. Oxford University Press.

Thelen, K. (2004). *How institutions evolve: The political economy of skills in Germany, Britain, the United States, and Japan*. Cambridge University Press.

Tinits, P. & Fey, C. F. (2022). The effects of timing and order of government support mechanisms for SME exports. *Management International Review, 62*, 285–323.

Tripsas, M., & Gavetti, G. (2000). Capabilities, cognition, and inertia: Evidence from digital imaging. *Strategic Management Journal, 21*(10/11), 1147–1161.

UFI (Global Association of the Exhibition Industry). (2021a). *Media release – UFI releases figures stating the impact of COVID-19 on the global exhibition economy in 2020*. Retrieved June 8, 2022, from https://www.ufi.org/wp-content/uploads/2021/02/MR_UFI-releases-figures-stating-the-impact-of-COVID-19_4february2021.pdf.

UFI (Global Association of the Exhibition Industry). (2021b). *Global recovery insights 2021: The road to recovery*. Retrieved June 8, 2022, from https://www.ufi.org/wpcontent/uploads/2022/04/UFI_Global_Recovery_Insights_2021.pdf.

UFI (Global Association of the Exhibition Industry). (2022). *Media Release. Global barometer provides updates on COVID-19 impact and 2022 perspectives for the exhibition industry*. Retrieved June 8, 2022, from https://www.ufi.org/mediarelease/ufi-global-barometer-provides-updates-on-covid-19-impact-and-2022-perspectives-for-the-exhibition-industry/

Yang, H., Steensma, H. K., & Phelps, C. (2010). Learning from what others have learned from you: The effects of knowledge spillovers on originating firms. *Academy of Management Journal, 53*(2), 371–389.

Appendices

Appendix A: Questionnaire

Inter-firm Interaction and Knowledge Flows at International Trade Fairs

---

1. How many years has your firm been attending international trade fairs?____________________

2. How many international trade fairs does your firm attend per year? ____________________

3. In which countries does your firm attend trade fairs? ____________________

4. Please rate the following capabilities that you consider to be central to the operations of your firm (i.e., this branch).

	Least Important — Most Important
(a) Innovation focus	1 ---- 2 ---- 3 ---- 4 ---- 5 ---- 6 ---- 7
(b) Low-cost focus	1 ---- 2 ---- 3 ---- 4 ---- 5 ---- 6 ---- 7
(c) Top-of-the-line production	1 ---- 2 ---- 3 ---- 4 ---- 5 ---- 6 ---- 7
(d) Strong R&D capabilities	1 ---- 2 ---- 3 ---- 4 ---- 5 ---- 6 ---- 7
(e) Marketing excellence	1 ---- 2 ---- 3 ---- 4 ---- 5 ---- 6 ---- 7
(f) Customized production	1 ---- 2 ---- 3 ---- 4 ---- 5 ---- 6 ---- 7
(g) Constantly changing product line	1 ---- 2 ---- 3 ---- 4 ---- 5 ---- 6 ---- 7

5. Do you look for potential partners at trade fairs? Yes ☐ No ☐ Sometimes ☐

R. Gibson, *International Trade Fairs and Inter-Firm Knowledge Flows*,

6. Please rate the following competencies that your firm looks for in potential partners or other firms at the trade fair.

	Least Important ... Most Important
(a) Innovation focus	1 ---- 2 ---- 3 ---- 4 ---- 5 ---- 6 ---- 7
(b) Low-cost focus	1 ---- 2 ---- 3 ---- 4 ---- 5 ---- 6 ---- 7
(c) Top-of-the-line production	1 ---- 2 ---- 3 ---- 4 ---- 5 ---- 6 ---- 7
(d) Strong R&D capabilities	1 ---- 2 ---- 3 ---- 4 ---- 5 ---- 6 ---- 7
(e) Marketing excellence	1 ---- 2 ---- 3 ---- 4 ---- 5 ---- 6 ---- 7
(f) Customized production	1 ---- 2 ---- 3 ---- 4 ---- 5 ---- 6 ---- 7
(g) Constantly changing product line	1 ---- 2 ---- 3 ---- 4 ---- 5 ---- 6 ---- 7

7. Does your firm engage in R&D collaborations with other firms? Yes ☐ No ☐

8. Has your firm merged with or acquired another firm in the past 3 years? Yes ☐ No ☐

9. Does your firm co-operate with external organizations in industry standard-setting? Yes ☐ No ☐

10. Where does your firm get knowledge from during the trade fair?

	Least Important ... Most Important
(a) Existing customers	1 ---- 2 ---- 3 ---- 4 ---- 5 ---- 6 ---- 7
(b) Potential customers	1 ---- 2 ---- 3 ---- 4 ---- 5 ---- 6 ---- 7
(c) Competitors	1 ---- 2 ---- 3 ---- 4 ---- 5 ---- 6 ---- 7
(d) Existing suppliers	1 ---- 2 ---- 3 ---- 4 ---- 5 ---- 6 ---- 7
(e) Potential suppliers	1 ---- 2 ---- 3 ---- 4 ---- 5 ---- 6 ---- 7
(f) Other partners in research/distribution	1 ---- 2 ---- 3 ---- 4 ---- 5 ---- 6 ---- 7

11. What types of knowledge does your firm get?

	Least Important ... Most Important
(a) Product-related	1 ---- 2 ---- 3 ---- 4 ---- 5 ---- 6 ---- 7
(b) Technology-related	1 ---- 2 ---- 3 ---- 4 ---- 5 ---- 6 ---- 7
(c) Market-related	1 ---- 2 ---- 3 ---- 4 ---- 5 ---- 6 ---- 7
(d) Industry-related	1 ---- 2 ---- 3 ---- 4 ---- 5 ---- 6 ---- 7
(e) Management-related	1 ---- 2 ---- 3 ---- 4 ---- 5 ---- 6 ---- 7

12. To whom does your firm provide knowledge?

	Least Important ... Most Important
(a) Existing customers	1 ---- 2 ---- 3 ---- 4 ---- 5 ---- 6 ---- 7
(b) Potential customers	1 ---- 2 ---- 3 ---- 4 ---- 5 ---- 6 ---- 7
(c) Competitors	1 ---- 2 ---- 3 ---- 4 ---- 5 ---- 6 ---- 7
(d) Existing suppliers	1 ---- 2 ---- 3 ---- 4 ---- 5 ---- 6 ---- 7
(e) Potential suppliers	1 ---- 2 ---- 3 ---- 4 ---- 5 ---- 6 ---- 7
(f) Other partners in research/distribution	1 ---- 2 ---- 3 ---- 4 ---- 5 ---- 6 ---- 7

13. What types of knowledge does your firm provide?

	Least Important ... Most Important
(a) Product-related	1 ---- 2 ---- 3 ---- 4 ---- 5 ---- 6 ---- 7
(b) Technology-related	1 ---- 2 ---- 3 ---- 4 ---- 5 ---- 6 ---- 7
(c) Market-related	1 ---- 2 ---- 3 ---- 4 ---- 5 ---- 6 ---- 7
(d) Industry-related	1 ---- 2 ---- 3 ---- 4 ---- 5 ---- 6 ---- 7
(e) Management-related	1 ---- 2 ---- 3 ---- 4 ---- 5 ---- 6 ---- 7

14. Please rate the following activities in which your firm participates in order to learn about the industry.

	Least Important ... Most Important
(a) Use of *regional* industry/association events	1 ---- 2 ---- 3 ---- 4 ---- 5 ---- 6 ---- 7
(b) Use of *national* industry/association events	1 ---- 2 ---- 3 ---- 4 ---- 5 ---- 6 ---- 7
(c) Use of specific industry websites	1 ---- 2 ---- 3 ---- 4 ---- 5 ---- 6 ---- 7
(d) Use of general social media *(e.g., blogs, Twitter, Facebook)*	1 ---- 2 ---- 3 ---- 4 ---- 5 ---- 6 ---- 7
(e) Attend web-casts	1 ---- 2 ---- 3 ---- 4 ---- 5 ---- 6 ---- 7
(f) Subscribe to Google alerts	1 ---- 2 ---- 3 ---- 4 ---- 5 ---- 6 ---- 7
(g) Attend industry conferences/conventions	1 ---- 2 ---- 3 ---- 4 ---- 5 ---- 6 ---- 7
(h) Attend international/national trade fairs	1 ---- 2 ---- 3 ---- 4 ---- 5 ---- 6 ---- 7
(i) Share professional information/advice over the phone	1 ---- 2 ---- 3 ---- 4 ---- 5 ---- 6 ---- 7
(j) Participate in lunch meetings with other firms/organizations	1 ---- 2 ---- 3 ---- 4 ---- 5 ---- 6 ---- 7
(k) Participate in professional organizations/associations	1 ---- 2 ---- 3 ---- 4 ---- 5 ---- 6 ---- 7

15. What is your firm's main product group? ____________________

16. What year was the firm first established? Headquarters: __________ This Branch: __________

17. How many employees does the firm have? Headquarters: __________ This Branch: __________

18. What is the firm's nationality (headquarters country)? ____________________

19. Are the operations of the firm mainly international or domestic? ☐ International ☐ Domestic

20. What are the firm's average annual sales? Headquarters: __________ This Branch: __________

21. What is your position in the firm? (e.g., sales, management) ____________________

22. What percentage of the firm's leading customers/clients is generated through trade fair participation? ____

23. What percentage of the firm's sales of new products is generated through participation at trade fairs? ____

24. What percentage of your marketing budget is devoted to trade fair participation per year? __________

25. Has your firm ever used government support to participate in a trade fair? Yes ☐ No ☐

Appendix B: Interview Guide

I. Trade fair participation

1. What are your main reasons/objectives for participating at this trade fair?
2. Has your firm attended international trade fairs before?

   – *Since when? How many times per year?*

3. What distinguishes this trade fair from others in the US, Canada, Europe?
4. What is the particular focus of this fair?

   – *How important is the presentation of innovations/state-of-the-art technologies?*

II. Problem and solution identification

5. Who typically presents your firm during a trade fair?

   – *What are the specific tasks of these representatives?*

6. Do you inform yourself about new developments/other exhibits during this trade fair?

   – *How? (Talking to visitors at exhibit, checking out other exhibits, evening meetings)*

7. How much time do you spend checking/visiting other exhibits?
8. Do you have a particular goal in looking through the exhibits; what do you look for?

   – *Do you exactly know what you need; how do you go ahead searching?*
   – *Do you have a problem/bottleneck that you are trying to solve?*
   – *How do you know when you have found something interesting/useful?*
   – *How do you distinguish something that is interesting from something that is not?*

III. Network generation

9. Who do you interact with during trade fairs?

   – *List according to importance/frequency.*

10. Do you systematically look for new partners during this trade fair?

    – *In which area: distribution, production, projects?*
    – *How do you go about finding partners?*

11. How important are trade fairs compared to other ways of acquiring new partners?
12. What was the most notable outcome of any trade fair you have been to in recent years?

    – *Explain: when/where did this happen?*

IV. **Additional information**

13. How important are trade fairs compared to other ways of meeting customers?
14. Has the importance of this trade fair changed in recent years?
    - *How? What about other trade fairs?*
15. Has your firm ever used government support to participate in trade fairs?
    - *If not, why? If so, what type of program did you use?*
16. What do you expect to learn from this trade fair that is useful/important for you?
17. Firm characteristics: age, number of employees, sales, country of origin

Appendix C: Descriptive Results of Questionnaire

Below are the aggregate results for U.S. and German firms that completed the self-administered questionnaire in both online and paper formats.

Questionnaire Question	Descriptive Results	Mean (Standard Deviation)	Sample Size
Firm characteristics			
Year firm was first established	Earliest: 1815	1973 (34.8)	155
	Latest: 2010		
Number of employees firm has	Min.: 0	1048 (4941)	145
	Max.: 53,000		
Firm nationality	U.S.: 105 (64%)	–	165
	Germany: 60 (36%)		
Firm sector	Machinery: 61 (37%)	–	165
	Electronics: 104 (63%)		
Jurisdiction of operations for firm	Domestic: 80 (49%)	–	164
	International: 84 (51%)		
Firm activities			
Is the firm engaged in research and development	No: 74 (46%)	–	161
	Yes: 87 (54%)		
Has the firm merged or acquired another firm in the last 3 years	No: 114 (69%)	–	164
	Yes: 50 (31%)		
Has the firm co-operated with external organizations in industry standard-setting	No: 77 (48%)	–	160
	Yes: 83 (52%)		
Firm trade fair experiences			
Has the firm ever used government support to participate in trade fairs	No: 125 (83%)	–	150
	Yes: 25 (17%)		
Number of years firm has been attending international trade fairs	Min.: 0	20.7 (20.6)	163
	Max.: 120		
Number of trade fairs the firm attends/year	Min.: 1	7 (12.3)	160
	Max.: 100		
Does firm look for potential partners at trade fairs	No: 19 (12%)	–	163
	Sometimes: 44 (27%)		
	Yes: 100 (61%)		

(continued)

Questionnaire Question	Descriptive Results	Mean (Standard Deviation)	Sample Size
Trade fair activities: Firm capabilities central to firm (1–7 scale; 1 = least important, 7 = most important)			
Innovation focus	–	5.8 (1.2)	163
Low-cost focus	–	4.0 (1.7)	163
Top-of-the-line production	–	5.4 (1.4)	160
Strong R&D capabilities	–	5.4 (1.4)	162
Marketing excellence	–	5.4 (1.3)	164
Customized production	–	5.2 (1.7)	161
Constantly changing product line	–	4.3 (1.8)	163
Trade fair activities: Capabilities look for in potential partners (1–7 scale; 1 = least important, 7 = most important)			
Innovation focus	–	5.3 (1.4)	129
Low-cost focus	–	4.1 (1.6)	139
Top-of-the-line production	–	5.0 (1.5)	130
Strong R&D capabilities	–	4.8 (1.5)	136
Marketing excellence	–	5.0 (1.6)	139
Customized production	–	4.8 (1.6)	130
Constantly changing product line	–	4.3 (1.7)	133
Trade fair: Where firm gets knowledge (1–7 scale; 1 = least important, 7 = most important)			
Existing customers	–	5.7 (1.3)	163
Potential customers	–	5.6 (1.3)	162
Competitors	–	4.7 (1.4)	161
Existing suppliers	–	4.8 (1.6)	162
Potential suppliers	–	4.6 (1.6)	161
Other partners in research/distribution	–	4.6 (1.7)	156
Trade fairs: Types of knowledge firm gets (1–7 scale; 1 = least important, 7 = most important)			
Product-related	–	5.7 (1.3)	161
Technology-related	–	5.5 (1.3)	163
Market-related	–	5.6 (1.1)	162
Industry-related	–	5.5 (1.2)	160
Management-related	–	4.0 (1.6)	163
Trade fairs: Whom does firm provide knowledge to (1–7 scale; 1 = least important, 7 = most important)			
Existing customers	–	6.1 (1.1)	161
Potential customers	–	6.1 (1.1)	160
Competitors	–	2.7 (1.6)	157
Existing suppliers	–	4.6 (1.6)	156
Potential suppliers	–	4.4 (1.7)	156
Other partners in research/distribution	–	4.4 (1.7)	153

(continued)

Questionnaire Question	Descriptive Results	Mean (Standard Deviation)	Sample Size
Trade fairs: Type of knowledge firm provides (1–7 scale; 1 = least important, 7 = most important)			
Product-related	–	6.2 (1.0)	162
Technology-related	–	5.5 (1.4)	162
Market-related	–	4.8 (1.5)	162
Industry-related	–	4.8 (1.5)	161
Management-related	–	3.4 (1.7)	161
Activities firm participates in to learn about the industry (1–7 scale; 1 = least important, 7 = most important)			
Use of regional industry/association events		4.1 (1.7)	156
Use of national industry/association events		4.8 (1.7)	157
Use of specific industry websites		5.0 (1.5)	157
Use of general social media		3.5 (2.0)	156
Attend web-casts		3.1 (1.6)	152
Subscribe to Google alerts		3.5 (1.9)	152
Attend industry conferences/conventions		4.7 (1.7)	154
Attend international/national trade fairs		5.5 (1.4)	155
Share professional information/advice over the phone		4.4 (1.6)	154
Participate in lunch meetings with other firms/ organizations		3.5 (1.7)	153
Participate in professional organizations/ associations		4.0 (1.8)	154
Questionnaire mode			
Type of survey filled out	Paper: 111 (67%) Online: 54 (33%)	–	165

Appendix D: Description of Variables

Dependent Variable: The dependent variable in this study is the nature of inter-firm interaction processes in relation to the different institutional arrangements of capitalist varieties – namely, CMEs, such as Germany, and LMEs, like the U.S.

Independent Variables: This study recognizes the multiplicity of knowledge flows and, thus, focuses on a number of different aspects of such flows. In the quantitative analysis, four aspects of knowledge flows are explored:

1. Where does the firm acquire knowledge during the trade fair (e.g., customers, competitors, suppliers)?
2. What types of knowledge does the firm get (e.g., product-related, technology-related)?
3. To whom does the firm provide knowledge at the trade fair (e.g., customers, competitors)?
4. What types of knowledge does it provide (e.g., industry-related, market-related)?

The qualitative portion of the analysis examines the following aspects of knowledge flows:

1. Do firms actually search for knowledge at trade fairs?
2. What exactly do they search for?
3. How do they search?

Control Variables: The quantitative portion of the study includes a number of control variables. These include: firm size; firm age; the sector in which the firm operates; use of government support to participate in the trade fair; whether the firm engages in R&D collaborations with other firms; whether the firm has merged with another firm in the last 3 years; and whether the firm co-operates with external organizations in industry standard-setting. Taking into account a variety of firm characteristics is important in enhancing confidence that the effects of country, when discovered, are robust. Aside from firm size, age, and sector, all of the control variables are dichotomous questions with yes/no answers.

- *Firm size* is defined in terms of number of employees and is distinguished into two categories: (i) small; (ii) medium-sized and large (ML). Small firms are those with 100 or fewer employees; ML firms are those with more than 100 employees.
- *Firm age* is placed into two categories: (i) post-globalization; (ii) pre-globalization. Post-globalization firms are those that were established in 1990 or later, corresponding to the beginning of the globalization period. Pre-globalization firms are those that were established prior to 1990.
- *Firm sector*. The quantitative portion of the analysis compares the electronics and machinery sectors, whereas the two broad sectors compared in the qualitative portion of the analysis are design-focused and technology-focused trade fairs.

- *International versus domestic focus*. For the quantitative analysis, firms were asked to identify whether the operations of their firm were mainly international or domestic. Given that the core thesis of the VoC approach revolves around the specific institutional arrangements of the national context in which firms operate, taking into account the domestic versus international orientation of the firm's central operations is important in determining the actual role played by the institutional setting of the national-state.
- *Inter-firm R&D collaborations*. For the quantitative analysis, firms were asked to respond "yes" or "no" to the following question: Does your firm engage in R&D collaborations with other firms? Distinctive national patterns in R&D strategies have long been noted in the literature (Pauly & Reich, 1997). Such patterns can include differences in the location of R&D facilities (e.g., at home or abroad), the nature of R&D activities, as well as R&D spending patterns. With respect to inter-firm R&D collaborations, studies have shown that such arrangements offer a number of critical advantages when it comes to learning, innovation, and knowledge processes. These include providing firms with more timely access to new combinations of knowledge (Grant & Baden-Fuller, 2004; Linnarson, 2005), supplying new technological and learning opportunities, and accelerating product development (Gilsing et al., 2007; Quintana-García & Benavides-Velasco, 2011). While distinct from the types of inter-firm interactions and searches that take place at trade fairs, inter-firm R&D collaborations across political-economic settings can play an important role in shaping the behaviour of firms with regards to the types of information sources they consider most relevant or the types of core competencies they see as most important to focus on – for both themselves and future partners. For these reasons, this variable has been included in the study.
- *Merged with or acquired another firm*. In the quantitative study, firms were asked if they had (yes/no) merged with or acquired another firm in the past 3 years. Similar to the argument regarding the international versus domestic focus of a firm's operations, the question of whether firms have acquired or merged with another firm may be relevant from a VoC perspective. We know, for example, that the market-based nature of inter-firm relations in LMEs enables easier entry into distant sectors or technological fields through the acquisition of already established firms (Boschma & Capone, 2015). While such mergers or acquisitions can involve firms that are similar to one another in terms of their core competencies or technological specializations, as noted above, they can also involve the acquisition of firms that are fundamentally different. This latter scenario could lead to changes in the internal structure of the firm and influence its behaviour in relation to patterns of search and interaction at trade fairs. As such, it is an important variable to take into account.

- *Industry standard-setting.* The survey developed for the quantitative analysis also asked firms to respond "yes" or "no" to the following question: Does your firm co-operate with external organizations in industry standard-setting? Such collaborations could have a bearing on the technological and production pathways firms choose to follow. For this reason, it is an important variable to include in the analysis.
- *Government support for trade fair participation.* Finally, the quantitative analysis looked at whether exhibiting firms had ever used government support to participate in the trade fair. This was posed as a yes/no question. It is not entirely clear whether or how financial support from the government to participate in the trade fair might shape the firm's behaviour at these events. However, this variable seemed potentially valuable given that the types of firms that receive government support might differ in important and systematic ways from those that do not. Moreover, firms that receive government support for trade fair participation are sometimes situated in designated areas of the exhibition hall or are included in specific country pavilions. Such distinctions and placement could potentially influence how and with whom firms interact during the trade fair.

References

Boschma, R., & Capone, G. (2015). Institutions and diversification: Related versus unrelated diversification in a varieties of capitalism framework. *Research Policy, 44*, 1902–1914.

Gilsing, V., Lemmens, C. E., & Duysters, G. (2007). Strategic alliance networks and innovation: A deterministic and voluntaristic view combined. *Technology Analysis and Strategic Management, 19*(2), 227–249.

Grant, R. M., & Baden-Fuller, C. (2004). A knowledge accessing theory of strategic alliances. *Journal of Management Studies, 41*, 61–84.

Linnarson, H. (2005). Patterns of alignment in alliance structure and innovation. *Technology Analysis & Strategic Management, 17*(2), 161–181.

Pauly, L. W., & Reich, S. (1997). National structures and multinational corporate behavior: Enduring differences in the age of globalization. *International Organization, 51*, 1–30.

Quintana-García, C., & Benavides-Velasco, C. A. (2011). Knowledge organisation in R&D alliances: Its impact on product innovation. *Technology Analysis and Strategic Management, 23*(10), 1047–1061.